Polars Cookbook

Over 60 practical recipes to transform, manipulate,
and analyze your data using Python Polars 1.x

Yuki Kakegawa

Polars Cookbook

Group Product Manager: Apeksha Shetty
Publishing Product Manager: Deepesh Patel
Book Project Manager: Farheen Fathima and Urvi Sharma
Senior Editor: Nazia Shaikh
Technical Editor: Kavyashree K S
Copy Editor: Safis Editing
Proofreader: Nazia Shaikh
Indexer: Pratik Shirodkar
Production Designers: Aparna Bhagat, Shankar Kalbhor, and Prafulla Nikalje
Senior DevRel Marketing Coordinator: Nivedita Singh

First published: August 2024

Production reference: 2020924

Published by Packt Publishing Ltd.
Grosvenor House
11 St Paul's Square
Birmingham
B3 1RB, UK

ISBN 978-1-80512-115-2

www.packtpub.com

First and foremost, I'm forever grateful for my wife, who encouraged me to go on this endeavor and supported me throughout the process. Without her support and sacrifice, I couldn't have written this book, let alone build my career.

Second, a big thanks to the Packt team, who ensured the quality and timeline of the book.

Third, I'd like to thank the author of Polars, Ritchie Vink, and other contributors who made Polars come to life and continue to develop it.

Finally, I'd like to express my gratitude to you, the readers. Thank you for reading my book.

– **Yuki Kakegawa**

Foreword

"Came for the speed, stayed for the syntax"

That's a common refrain among Polars enthusiasts. Indeed, the Polars API is truly beautiful: not only does it make for very readable code, but it also allows you to express complex aggregations that just aren't expressible with the pandas API.

Yuki has been a long-time fan of Polars. He has professional experience as a consultant. It's great to see him pair these together to produce a cookbook of practical recipes that you can use to solve real problems.

When should you use Polars? I think the best time is when you're starting a new project. Porting pandas code to Polars is certainly possible, but it's not necessarily easy. If you try thinking in Polars at the start of a new project, you'll likely surprise yourself with how expressive its API truly is, you'll use it idiomatically, and you'll make full use of its amazing features.

I'm sure you'll love learning about Polars whilst reading this book. And when you start your next data science project - please join the Polars Discord to say hello! Would love to hear about your experience!

Marco Gorelli

Polars and Pandas Contributor | Senior Software Engineer, Quansight

Contributors

About the author

Yuki Kakegawa is a data analytics professional with a background in computer science. Yuki has worked in the data space for the past several years, most of which has been spent in consulting, focusing on data engineering, analytics, and business intelligence. His clients are from various industries, such as healthcare, education, insurance, and private equity. He has worked with various companies, including Microsoft and Stanford Health Care, to name a couple.

He also runs Orem Data, a data analytics consultancy that helps companies improve their existing data and analytics infrastructure.

Aside from work, Yuki enjoys playing baseball and softball with his wife and friends.

About the reviewer

Mihai Gurău is an analytics and data professional with over eight years of experience, focusing on the "why?" behind analytics to drive meaningful action. In the airline industry, he has helped build bespoke revenue management decision support tools. His process mining implementation work effectively melded analytics with enterprise IT systems for process discovery and improvement. Nowadays, he contributes to fine-tuning product analytics and building robust data platform components for map-making and connected products and services. Beyond his professional pursuits, Mihai enjoys watersports and tries to keep abreast of relevant advancements in data and analytics engineering.

Table of Contents

2

Reading and Writing Files   37

3

An Introduction to Data Analysis in Python Polars   67

4

Data Transformation Techniques   95

5

Handling Missing Data   131

6

Performing String Manipulations   149

7

Working with Nested Data Structures   175

8

Reshaping and Tidying Data   211

11

Working with Common Cloud Data Sources   287

12

Testing and Debugging in Polars   337

Preface

Polars is a lightning-fast DataFrame library for transforming and manipulating data. It is written in Rust from scratch and designed for efficient parallelism. Polars is available in Rust, Python, R, and Node.js. Polars has been gaining popularity in the world of data analysis, science, and engineering with its high performance and well-designed API.

However, as Polars is still new, there are not many resources available for you to learn how to use Polars to its full potential. The good news is that this book is designed to fill that gap.

This book contains about 60 recipes that teach you how to use various data transformation and manipulation techniques that you can apply across common data problems. You will unlock the power of Polars by going through the book while keeping the practical recipes on hand to refer to when needed.

Who this book is for

This book is for data analysts, data scientists, and data engineers who want to learn to use Polars in their workflows. You should have working knowledge of the Python programming language. Preferably, you have experience of working with a DataFrame library such as pandas or PySpark. If you're looking to improve your data pipelines or data science/analysis workflows with Python Polars, this book is ideal to get you started and be the guide in your journey.

Another audience that may be a good fit for this book is those who know Python, but have not used any DataFrame libraries before. The book is structured so that it starts with basics and fundamental concepts that are important across DataFrame libraries. The complexity of the concepts covered then increases as you progress through the later chapters. Throughout the book, you'll develop practical skills in using Polars as well as gaining a general understanding of the mechanics of a DataFrame library.

What this book covers

Chapter 1, Getting Started with Python Polars, introduces you to Polars' unique features and fundamental operations. This chapter guides you through the basic workings of DataFrames, LazyFrames, Series, and Expressions, getting you started on your journey with Python Polars.

Chapter 2, Reading and Writing Files, teaches you how to read from and write to various types of files and databases.

Chapter 3, An Introduction to Data Analysis in Python Polars, walks you through the application of common data exploration tasks. This includes, but is not limited to, inspecting the data, generating summary statistics, casting data types, removing duplicate values, and visualizing data.

Chapter 4, Data Transformation Techniques, helps you learn to apply aggregations, window functions, and **user-defined functions** (**UDFs**) in your data transformation pipelines. This chapter also introduces you how to use SQL in Polars.

Chapter 5, Handling Missing Data, covers ways to identify and handle missing values.

Chapter 6, Performing String Manipulations, helps you develop your skills in manipulating strings using Polars' built-in methods.

Chapter 7, Working with Nested Data Structures, teaches you how to work with nested data structures such as lists and structs.

Chapter 8, Reshaping and Tidying Data, introduces you to ways to transform your data from a wide format to a long format and vice versa. This chapter also covers how you can combine or concatenate DataFrames.

Chapter 9, Time Series Analysis, focuses on time series analysis techniques such as rolling calculations and resampling methods. You'll also learn to work with time-related attributes and how to build a simple time series forecasting model.

Chapter 10, Interoperability with Other Python Libraries, covers how Polars can interoperate with other Python libraries such as pandas, NumPy, PyArrow, and DuckDB.

Chapter 11, Working with Common Cloud Data Sources, gets you started on working with popular cloud object storage systems and databases.

Chapter 12, Testing and Debugging in Polars, teaches you how to use Polars' built-in testing methods/functions and how to debug and troubleshoot your code using libraries such as `pytest` and `cuallee`.

To get the most out of this book

You will need a version of Python installed on your computer. All code examples have been tested using Python 3.11 on macOS. However, they should work with future version releases, too. Please note that Polars requires Python version 3.8 or higher.

Software/hardware covered in the book	OS requirements
Python >= 3.8	Windows, macOS, or Linux

In any Python project, it is recommended that you use a virtual environment to isolate the dependencies of your different projects. This allows you to install libraries into an environment without affecting the global Python installation or any other environments.

The use of a virtual environment is not required, however, you can easily create one with venv, which comes as part of the Python standard library. The following code creates a virtual environment:

```
$ python -m venv .venv
```

If you're on macOS or Linux, the following code activates the virtual environment:

```
$ source .venv/bin/activate
```

You can use the following code for Windows:

```
$ .\.venv\Scripts\activate
```

There are several methods to install Polars and other Python libraries, however, in this book, we'll be using pip, a popular package manager for Python.

You can either install the necessary libraries all at once or separately as you go. If you want to take the former approach, run the following code to install all the dependencies included in the requirements.txt file:

```
$ pip install -r requirements.txt
```

If you want to take the latter approach, you'll first need to install Jupyter Notebook since all the code examples are written in Jupyter notebooks, and the book assumes you will be running the code using Jupyter Notebook.

Install JupyterLab with pip:

```
$ pip install notebook
```

Launch the notebook as follows:

```
$ jupyter notebook
```

Alternatively, you can install JupyterLab with pip with the following command:

```
$ pip install jupyterlab
```

Once installed, you can launch JupyterLab with the following command:

```
$ jupyter lab
```

You can also refer to the instructions on the Jupyter website: https://jupyter.org/install.

Note that if you're using an **Integrated Development Environment (IDE)** such as **Visual Studio Code (VSCode)**, you may not need to launch the notebook. Instead, you can open it directly in VSCode once you have installed the notebook/JupyterLab via pip and necessary VSCode extensions as required.

Several other libraries will be used throughout the book, including the Polars library. In each chapter and recipe, you'll be instructed to install these when necessary; however, it may be good to install the required libraries beforehand.

Use the following command to install Polars:

```
$ pip install polars
```

You can specify optional dependencies in brackets, as in the following command:

```
$ pip install polars[pyarrow, pandas]
```

To install all the optional dependencies of Polars, specify the term `all` in the brackets:

```
$ pip install polars[all]
```

Also, while this most likely won't be an issue, it is worth noting that your Polars should be version >= 1.0.0. That's the version all the code in this book was tested on.

Now that you have set up your Python environment and have installed the dependencies, you're ready to dive into the book contents.

If you are using the digital version of this book, we advise you to type the code yourself or access the code via the GitHub repository (link available in the next section). Doing so will help you avoid any potential errors related to the copying and pasting of code.

As Polars is growing rapidly day by day, it's possible that some of the code snippets or methods and functions used in the book will be outdated by the time you're reading it. We'll do our best to note the new features and deprecations in the chapter notebooks, however, please feel free to contact us by sending an email or creating an issue in the GitHub repo.

Download the example code files

You can download the example code files for this book from GitHub at https://github.com/ PacktPublishing/Polars-Cookbook/. If there's an update to the code, it will be updated on the existing GitHub repository.

We also have other code bundles from our rich catalog of books and videos available at https:// github.com/PacktPublishing/. Check them out!

Conventions used

There are a number of text conventions used throughout this book.

`Code in text`: Indicates code words in text, database table names, folder names, filenames, file extensions, pathnames, dummy URLs, user input, and Twitter handles. Here is an example: "`pl. read_csv()` is one of the common ways to read data into a DataFrame."

A block of code is set as follows:

```
import numpy as np

numpy_arr = np.array([[1,1,1], [2,2,2]])
df = pl.from_numpy(numpy_arr, schema={'ones': pl.Float32, 'twos':
pl.Int8}, orient='col')
df.head()
```

Any command-line input or output is written as follows:

```
$ pip install polars
```

> **Tips or important notes**
> Appear like this.

Sections

In this book, you will find several headings that appear frequently (*Getting ready*, *How to do it...*, *How it works...*, *There's more...*, and *See also*).

To give clear instructions on how to complete a recipe, use these sections as follows:

Getting ready

This section tells you what to expect in the recipe and describes how to set up any software or any preliminary settings required for the recipe.

How to do it...

This section contains the steps required to follow the recipe.

How it works...

This section usually consists of a detailed explanation of what happened in the previous section.

There's more...

This section consists of additional information about the recipe in order to make you more knowledgeable about the recipe.

See also

This section provides helpful links to other useful information for the recipe.

Get in touch

Feedback from our readers is always welcome.

General feedback: If you have questions about any aspect of this book, mention the book title in the subject of your message and email us at customercare@packtpub.com.

Errata: Although we have taken every care to ensure the accuracy of our content, mistakes do happen. If you have found a mistake in this book, we would be grateful if you would report this to us. Please visit www.packtpub.com/support/errata, select your book, click on the **Errata Submission Form** link, and enter the details.

Piracy: If you come across any illegal copies of our works in any form on the Internet, we would be grateful if you would provide us with the location address or website name. Please contact us at copyright@packt.com with a link to the material.

If you are interested in becoming an author: If there is a topic that you have expertise in and you are interested in either writing or contributing to a book, please visit authors.packtpub.com.

Share Your Thoughts

Once you've read *Polars Cookbook*, we'd love to hear your thoughts! Scan the QR code below to go straight to the Amazon review page for this book and share your feedback.

https://packt.link/r/1-805-12115-4

Your review is important to us and the tech community and will help us make sure we're delivering excellent quality content.

Download a free PDF copy of this book

Thanks for purchasing this book!

Do you like to read on the go but are unable to carry your print books everywhere?

Is your eBook purchase not compatible with the device of your choice?

Don't worry, now with every Packt book you get a DRM-free PDF version of that book at no cost.

Read anywhere, any place, on any device. Search, copy, and paste code from your favorite technical books directly into your application.

The perks don't stop there, you can get exclusive access to discounts, newsletters, and great free content in your inbox daily

Follow these simple steps to get the benefits:

1. Scan the QR code or visit the link below

https://packt.link/free-ebook/978-1-80512-115-2

2. Submit your proof of purchase
3. That's it! We'll send your free PDF and other benefits to your email directly

1

Getting Started with Python Polars

This chapter will look at the fundamentals of Python Polars. We will learn some of the key features of Polars at a high level in order to understand why Polars is fast and efficient for processing data. We will also cover how to apply basic operations on DataFrame, Series, and LazyFrame utilizing Polars expressions. These are all essential bits of knowledge and techniques to start utilizing Polars in your data workflows.

This chapter contains the following recipes:

- Introducing key features in Polars
- The Polars DataFrame
- Polars Series
- The Polars LazyFrame
- Selecting columns and filtering data
- Creating, modifying, and deleting columns
- Understanding method chaining
- Processing larger-than-RAM datasets

After going through all of these, you'll have a good understanding of what makes Polars unique, as well as how to apply essential data operations in Polars.

Technical requirements

As explained in the *Preface*, you'll need to set up your Python environment and install and import the Polars library. Here's how to install the Polars library using `pip`:

```
>>> pip install polars
```

If you want to install all the optional dependencies, you'll need to use the following:

```
>>> pip install 'polars[all]'
```

If you want to install specific optional dependencies, you'll use the following:

```
>>> pip install 'polars[pyarrow, pandas]'
```

Here's a line of code to import the Python Polars library:

```
import polars as pl
```

You can find the code and dataset from this chapter along with datasets used in the GitHub repository here: `https://github.com/PacktPublishing/Polars-Cookbook`.

In addition to Polars, you will need to install the Graphviz library, which is required to visually inspect the query plan:

```
>>> pip install graphviz
```

You will also need to install the Graphviz package on your machine. Please refer to this website for how to install the package on your chosen OS: `https://graphviz.org/download/`.

I installed it on my Mac using Homebrew with the following command:

```
>>> brew install graphviz
```

For Windows users, the simplified steps are as follows:

1. Select whether you want to install the 32-bit or the 64-bit version of Graphviz.
2. Visit the download location at `https://gitlab.com/graphviz/graphviz/-/releases`.
3. Download the 32-bit or 64-bit exe file:
    I. The 32-bit `.exe` file: `https://gitlab.com/graphviz/graphviz/-/package_files/6164165/download`
    II. The 64-bit `.exe` file: `https://gitlab.com/graphviz/graphviz/-/package_files/6164164/download`

Please refer to these instructions for a more detailed explanation of how to install Graphviz on Windows: `https://forum.graphviz.org/t/new-simplified-installation-procedure-on-windows/224`.

You can find more information about Graphviz in general here: `https://graphviz.readthedocs.io/en/stable/`.

Introducing key features in Polars

Polars is a blazingly fast DataFrame library that allows you to manipulate and transform your structured data. It is designed to work on a single machine utilizing all the available CPUs.

There are many other DataFrame libraries in Python including pandas and PySpark. Polars is one of the newest DataFrame libraries. It is performant and it has been gaining popularity at lightning speed.

A DataFrame is a two-dimensional structure that contains one or more Series. A Series is a one-dimensional structure, array, or list. You can think of a DataFrame as a table and a Series as a column. However, Polars is so much more. There are concepts and features that make Polars a fast and high-performant DataFrame library. It's good to have at least some level of understanding of these key features to maximize your learning and effective use of Polars.

At a high level, these are the key features that make Polars unique:

- Speed and efficiency
- Expressions
- The lazy API

Speed and efficiency

We know that Polars is fast and efficient. But what has contributed to making Polars the way it is today? There are a few main components that contribute to its speed and efficiency:

- The Rust programming language
- The Apache Arrow columnar format
- The lazy API

Polars is written in Rust, a low-level programming language that gives a similar level of performance and full control over memory as C/C++. Because of the support for concurrency in Rust, Polars can execute many operations in parallel, utilizing all the CPUs available on your machine without any configuration. We call that **embarrassingly parallel execution**.

Also, Polars is based on Apache Arrow's columnar memory format. That means that Polars can not only utilize the optimization of columnar memory but also share data between other Arrow-based tools for free without copying the data every time (using pointers to the original data, eliminating the need to copy data around).

Finally, the lazy API makes Polars even faster and more efficient by implementing several other query optimizations. We'll cover that in a second under *The lazy API*.

These core components have essentially made it possible to implement the features that make Polars so fast and efficient.

Expressions

Expressions are what makes Polars's syntax readable and easy to use. Its expressive syntax allows you to write complex logic in an organized, efficient fashion. Simply put, an expression takes a Series as an input and gives back a Series as an output (think of a Series like a column in a table or DataFrame). You can combine multiple expressions to build complex queries. This chain of expressions is the essence that makes your query even more powerful.

An expression takes a Series and gives back a Series as shown in the following diagram:

Figure 1.1 – The Polars expressions mechanism

Multiple expressions work on a Series one after another as shown in the following diagram:

Figure 1.2 – Chained Polars expressions

As it relates to expressions, **context** is an important concept. A context is essentially the environment in which an expression is evaluated. In other words, expressions can be used when you expose them within a context. Of the contexts you have access to in Polars, these are the three main ones:

- Selection
- Filtering
- Group by/aggregation

We'll look at specific examples and use cases of how you can utilize expressions in these contexts throughout the book. You'll unlock the power of Polars as you learn to understand and use expressions extensively in your code.

Expressions are part of the clean and simple Polars API. This provides you with better ergonomics and usability for building your data transformation logic in Polars.

The lazy API

The lazy API makes Polars even faster and more efficient by applying additional optimizations such as predicate pushdown and projection pushdown. It also optimizes the query plan automatically, meaning that Polars figures out the most optimal way of executing your query. You can access the lazy API by using LazyFrame, which is a different variation of DataFrame.

The lazy API uses **lazy evaluation**, which is a strategy that involves delaying the evaluation of an expression until the resulting value is needed. With the lazy API, Polars processes your query end-to-end instead of processing it one operation at a time. You can see the full list of optimizations available with the lazy API in the Polars user guide here: `https://pola-rs.github.io/polars/user-guide/lazy/optimizations/`.

One other feature that's available in the lazy API is streaming processing or the streaming API. It allows you to process data that's larger than the amount of memory available on your machine. For example, if you have 16 GB of RAM on your laptop, you may be able to process 50 GB of data.

However, it's good to keep in mind that there is a limitation. Although this larger-than-RAM processing feature is available on many of the operations, not all operations are available (as of the time of authoring the book).

> **Note**
> Eager evaluation is another evaluation strategy in which an expression is evaluated as soon as it is called. The Polars DataFrame and other DataFrame libraries like pandas use it by default.

See also

To learn more about how Python Polars works, including its optimizations and mechanics, please refer to these resources:

- `https://pola-rs.github.io/polars/`
- `https://pola-rs.github.io/polars/user-guide/lazy/optimizations/`
- `https://blog.jetbrains.com/dataspell/2023/08/polars-vs-pandas-what-s-the-difference/`

- Ritchie Vink Polars; done the fast, now the scale PyCon 2023 - `https://www.youtube.com/watch?v=apuFzB4j2_E&list=LL&index=5`
- Polars, the fastest DataFrame library you never heard of - `https://www.youtube.com/watch?v=pzx99Mp52C8&list=LL&index=5`
- Polars, the Fastest Dataframe Library You Never Heard of. - Ritchie Vink | PyData Global 2021 - `https://www.youtube.com/watch?v=iwGIuGk5nCE`

The Polars DataFrame

DataFrame is the base component of Polars. It is worth learning its basics as you begin your journey in Polars. DataFrame is like a table with rows and columns. It's the fundamental structure that other Polars components are deeply interconnected with.

If you've used the pandas library before, you might be surprised to learn that Polars actually doesn't have a concept of an **index**. In pandas, an index is a series of labels that identify each row. It helps you select and align rows of your DataFrame. This is also different from the indexes you might see in SQL databases in that an index in pandas is not meant to apply for a faster data retrieval performance.

You might've found index in pandas useful, but I bet that they also gave you some headaches. Polars avoids the complexity that comes with index. If you'd like to learn more about the differences in concepts between pandas and Polars, you can look at this page in the Polars documentation: `https://pola-rs.github.io/polars/user-guide/migration/pandas`.

In this recipe, we'll cover some ways to create a Polars DataFrame, as well as useful methods to extract DataFrame attributes.

Getting ready

We'll use a dataset stored in this GitHub repo: `https://github.com/PacktPublishing/Polars-Cookbook/blob/main/data/titanic_dataset.csv`. Also, make sure that you import the Polars library at the beginning of your code:

```
Import polars as pl
```

How to do it...

We'll start by creating a DataFrame and exploring its attributes.:

1. Create a DataFrame from scratch with a Python dictionary as the input:

```
df = pl.DataFrame({
    'nums': [1,2,3,4,5],
    'letters': ['a','b','c','d','e']
})
df.head()
```

The preceding code will return the following output:

shape: (5, 2)

nums	letters
i64	str
1	"a"
2	"b"
3	"c"
4	"d"
5	"e"

Figure 1.3 – The output of an example DataFrame

2. Create a DataFrame by reading a `.csv` file. Then take a peek at the dataset:

```
df = pl.read_csv('../data/titanic_dataset.csv')
df.head()
```

The preceding code will return the following output:

shape: (5, 12)

PassengerId	Survived	Pclass	Name	Sex	Age	SibSp	Parch	Ticket	Fare	Cabin	Embarked
i64	i64	i64	str	str	f64	i64	i64	str	f64	str	str
1	0	3	"Braund, Mr. Ow...	"male"	22.0	1	0	"A/5 21171"	7.25	null	"S"
2	1	1	"Cumings, Mrs. ...	"female"	38.0	1	0	"PC 17599"	71.2833	"C85"	"C"
3	1	3	"Heikkinen, Mis...	"female"	26.0	0	0	"STON/O2. 31012...	7.925	null	"S"
4	1	1	"Futrelle, Mrs....	"female"	35.0	1	0	"113803"	53.1	"C123"	"S"
5	0	3	"Allen, Mr. Wil...	"male"	35.0	0	0	"373450"	8.05	null	"S"

Figure 1.4 – The first few rows of the titanic dataset

Explore DataFrame attributes. `.schemas` gives you the combination of each column name and data type in Python dictionary. You can get column names and data types in separate lists with `.columns` and `.dtypes`:

```
df.schema
```

The preceding code will return the following output:

```
>> Schema([('PassengerId', Int64), ('Survived', Int64),
('Pclass', Int64), ('Name', String), ('Sex', String), ('Age',
Float64), ('SibSp', Int64), ('Parch', Int64), ('Ticket',
String), ('Fare', Float64), ('Cabin', String), ('Embarked',
String)])
```

```
df.columns
```

The preceding code will return the following output:

```
>> ['PassengerId', 'Survived', 'Pclass', 'Name', 'Sex', 'Age',
'SibSp', 'Parch', 'Ticket', 'Fare', 'Cabin', 'Embarked']
```

```
df.dtypes
```

The preceding code will return the following output:

```
>> [Int64, Int64, Int64, String, String, Float64, Int64, Int64,
String, Float64, String, String]
```

You can get the height and width of your DataFrame with `.shape`. You can also get the height and width individually with `.height` and `.width` as well:

```
df.shape
```

The preceding code will return the following output:

```
>> (891, 12)
```

```
df.height
```

The preceding code will return the following output:

```
>> 891
```

```
df.width
```

The preceding code will return the following output:

```
>> 12
```

```
df.flags
```

The preceding code will return the following output:

```
>> {'PassengerId': {'SORTED_ASC': False, 'SORTED_DESC': False},
 'Survived': {'SORTED_ASC': False, 'SORTED_DESC': False},
 'Pclass': {'SORTED_ASC': False, 'SORTED_DESC': False},
 'Name': {'SORTED_ASC': False, 'SORTED_DESC': False},
 'Sex': {'SORTED_ASC': False, 'SORTED_DESC': False},
 'Age': {'SORTED_ASC': False, 'SORTED_DESC': False},
 'SibSp': {'SORTED_ASC': False, 'SORTED_DESC': False},
 'Parch': {'SORTED_ASC': False, 'SORTED_DESC': False},
 'Ticket': {'SORTED_ASC': False, 'SORTED_DESC': False},
 'Fare': {'SORTED_ASC': False, 'SORTED_DESC': False},
 'Cabin': {'SORTED_ASC': False, 'SORTED_DESC': False},
 'Embarked': {'SORTED_ASC': False, 'SORTED_DESC': False}}
```

How it works...

Within `pl.DataFrame()`, I have added a Python dictionary as the data source. Its keys are strings, and its values are lists. Data types are auto-inferred unless you specify the schema.

The `.head()` method is handy in your analysis workflow. It shows the first *n* rows, where *n* is the number of rows you specify. The default value of *n* is set to 5.

`pl.read_csv()` is one of the common ways to read data into a DataFrame. It involves specifying the path of the file you want to read. It has many parameters that help you load data efficiently, tailored to your use case. We'll cover the topic of reading and writing files in detail in the next chapter.

There's more...

The Polars DataFrame can take many forms of data as its source, such as Python dictionaries, the Polars Series, NumPy array, pandas DataFrame, and so on. You can even utilize functions like `pl.from_numpy()` and `pl.from_pandas()` to import data directly from other structures instead of using `pl.DataFrame()`.

Also, there are several parameters you can set when creating a DataFrame, including the schema. You can preset the schema of your dataset, or else it will be auto-inferred by Polars's engine:

```
import numpy as np

numpy_arr = np.array([[1,1,1], [2,2,2]])
df = pl.from_numpy(numpy_arr, schema={'ones': pl.Float32, 'twos':
pl.Int8}, orient='col')
df.head()
```

The preceding code will return the following output:

shape: (3, 2)

ones	twos
f32	i8
1.0	2
1.0	2
1.0	2

Figure 1.5 – A DataFrame created from a NumPy array

Both reading into a DataFrame and outputting to other structures such as pandas DataFrame and pyarrow.Table is possible. We'll cover that in *Chapter 10, Interoperability with Other Python Libraries*.

You can basically categorize the data types in Polars into five categories:

- Numeric
- String/categorical
- Date/time

- Nested
- Other (Boolean, Binary, and so forth)

We'll look at working with specific types of data throughout this book, but it's good to know what data types exist early on in the journey of learning about Polars.

You can see a complete list of data types on this Polars documentation page: `https://pola-rs.github.io/polars/py-polars/html/reference/datatypes.html`.

See also

Please refer to each section of the Polars documentation for additional information:

- `https://pola-rs.github.io/polars/py-polars/html/reference/dataframe`
- `https://pola-rs.github.io/polars/py-polars/html/reference/api/polars.read_csv.html`
- `https://pola-rs.github.io/polars/py-polars/html/reference/dataframe/attributes.html`
- `https://pola-rs.github.io/polars/py-polars/html/reference/functions.html`

Polars Series

Series is an important concept in a DataFrame library. A DataFrame is made up of one or more Series. A Series is like a list or array: it's a one-dimensional structure that stores a list of values. A Series is different than a list or array in Python in that a Series is viewed as a column in a table, containing the list of data points or values of a certain data type. Just like the Polars DataFrame, the Polars Series also has many built-in methods you can utilize for your data transformations. In this recipe, we'll cover the creation of Polars Series as well as how to inspect its attributes.

Getting ready

As usual, make that sure you import the Polars library at the beginning of your code if you haven't already:

```
import polars as pl
```

How to do it...

We'll first create a Series and explore its attributes.

1. Create a Series from scratch:

```
s = pl.Series('col', [1,2,3,4,5])
s.head()
```

The preceding code will return the following output:

shape: (5,)

col
i64
1
2
3
4
5

Figure 1.6 – Polars Series

2. Create a Series from a DataFrame with the .to_series() and .get_column() methods:

    I. First, let's convert a DataFrame to a Series with .to_series():

```
data = {'a': [1,2,3], 'b': [4,5,6]}
s_a = (
    pl.DataFrame(data)
    .to_series()
)
s_a.head()
```

The preceding code will return the following output:

shape: (3,)

a
i64
1
2
3

Figure 1.7 – A Series from a DataFrame

II.   By default, `.to_series()` returns the first column. You can specify the column by either index:

```
s_b = (
    pl.DataFrame(data)
    .to_series(1)
)
s_b.head()
```

III. When you want to retrieve a column for a Series, you can use `.get_columns()` instead:

```
s_b2 = (
    pl.DataFrame(data)
    .get_column('b')
)
s_b2.head()
```

The preceding code will return the following output:

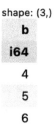

shape: (3,)

b
i64
4
5
6

Figure 1.8 – Different ways to extract a Series from a DataFrame

3. Display Series attributes:

I.   Get the length and width with `.shape`:

```
s.shape
```

The preceding code will return the following output:

```
>> (5,)
```

II.   Use `.name` to get the column name:

```
s.name
```

The preceding code will return the following output:

```
>> 'col'
```

III.   .dtype gives you the data type:

```
s.dtype
```

The preceding code will return the following output:

```
>> Int64
```

How it works...

The process of creating a Series and getting its attributes is similar to that of creating a DataFrame. There are many other methods that are common across DataFrame and Series. Knowing how to work with DataFrame means knowing how to work with Series and vice-versa.

There's more...

Just like DataFrame, Series can be converted between other structures such as a NumPy array and pandas Series. We won't get into details on that in this book, but we'll go over this for DataFrame later in the book in *Chapter 10, Interoperability with Other Python Libraries*.

See also

If you'd like to learn more, please visit Polars' documentation page: https://pola-rs.github. io/polars/py-polars/html/reference/series/index.html.

The Polars LazyFrame

One of the unique features that makes Polars even faster and more efficient is its lazy API. The lazy API uses lazy evaluation, a technique that delays the evaluation of an expression until its value is needed. That means your query is only executed when it's needed. This allows Polars to apply query optimizations because Polars can look at and execute multiple transformation steps at once by looking at the computation graph as a whole only when you tell it to do so. On the other hand, with eager evaluation (another evaluation strategy you'd use with DataFrame), you process data every time per expression. Essentially, lazy evaluation gives you more efficient ways to process your data.

You can access the Polars lazy API by using what we call LazyFrame. As explained earlier, LazyFrame allows for automatic query optimizations and larger-than-RAM processing.

LazyFrame is the proffered way of using Polars simply because it has more features and abilities to handle your data better. In this recipe, you'll learn how to create a LazyFrame as well as how to use useful methods and functions associated with LazyFrame.

How to do it...

We'll explore a LazyFrame by creating it first. Here are the steps:

1. Create a LazyFrame from scratch:

```
data = {'name': ['Sarah', 'Mike', 'Bob', 'Ashley']}
lf = pl.LazyFrame(data)
type(lf)
```

The preceding code will return the following output:

```
>> polars.lazyframe.frame.LazyFrame
```

2. Use the `.collect()` method to instruct Polars to process data:

```
lf.collect().head()
```

The preceding code will return the following output:

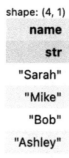

Figure 1.9 – LazyFrame output

3. Create a LazyFrame from a `.csv` file using the `.scan_csv()` method:

```
lf = pl.scan_csv('../data/titanic_dataset.csv')
lf.head().collect()
```

The preceding code will return the following output:

shape: (5, 12)

PassengerId	Survived	Pclass	Name	Sex	Age	SibSp	Parch	Ticket	Fare	Cabin	Embarked
i64	i64	i64	str	str	f64	i64	i64	str	f64	str	str
1	0	3	"Braund, Mr. Ow...	"male"	22.0	1	0	"A/5 21171"	7.25	null	"S"
2	1	1	"Cumings, Mrs. ...	"female"	38.0	1	0	"PC 17599"	71.2833	"C85"	"C"
3	1	3	"Heikkinen, Mis...	"female"	26.0	0	0	"STON/O2. 31012...	7.925	null	"S"
4	1	1	"Futrelle, Mrs....	"female"	35.0	1	0	"113803"	53.1	"C123"	"S"
5	0	3	"Allen, Mr. Wil...	"male"	35.0	0	0	"373450"	8.05	null	"S"

Figure 1.10 – The output of using .scan_csv()

4. Convert a LazyFrame from a DataFrame with the `.lazy()` method:

```
df = pl.read_csv('../data/titanic_dataset.csv')
df.lazy().head(3).collect()
```

The preceding code will return the following output:

shape: (3, 12)

PassengerId	Survived	Pclass	Name	Sex	Age	SibSp	Parch	Ticket	Fare	Cabin	Embarked
i64	i64	i64	str	str	f64	i64	i64	str	f64	str	str
1	0	3	"Braund, Mr. Ow...	"male"	22.0	1	0	"A/5 21171"	7.25	null	"S"
2	1	1	"Cumings, Mrs. ...	"female"	38.0	1	0	"PC 17599"	71.2833	"C85"	"C"
3	1	3	"Heikkinen, Mis...	"female"	26.0	0	0	"STON/O2. 31012...	7.925	null	"S"

Figure 1.11 – Convert a DataFrame into a LazyFrame

5. Show the schema and width of LazyFrame:

```
lf.collect_schema()
```

The preceding code will return the following output:

```
>> Schema([('PassengerId', Int64), ('Survived', Int64),
('Pclass', Int64), ('Name', String), ('Sex', String), ('Age',
Float64), ('SibSp', Int64), ('Parch', Int64), ('Ticket',
String), ('Fare', Float64), ('Cabin', String), ('Embarked',
String)])
```

```
lf.collect_schema().len()
```

The preceding code will return the following output:

```
>> 12
```

How it works...

The structure of LazyFrame is the same as that of DataFrame, but LazyFrame doesn't process your query until it's told to do so using `.collect()`. You can use this to trigger the execution of the computation graph or query of a LazyFrame. This operation materializes a LazyFrame into a DataFrame.

> **Note**
>
> You should keep in mind that some operations that are available in DataFrame are not available in LazyFrame (such as `.pivot()`). These operations require Polars to know the whole structure of the data, which LazyFrame is not capable of handling. However, once you use `.collect()` to materialize a DataFrame, you'll be able to use all the available DataFrame methods on it.

The way in which you create a LazyFrame is similar to the method for creating a DataFrame. After you have created a LazyFrame, and once it's been materialized with `.collect()`, LazyFrame is converted to DataFrame. That's why you can call `.head()` on it after calling `.collect()`.

> **Note**
>
> You may be aware of the `.fetch()` method that was available until Polars version 0.20.31. While it was useful for debugging purposes, there were some gotchas that were confusing to users. Since Polars version 1.0.0, this method is deprecated. It's still available as `._fetch()` for development purposes.

You will notice that when you read a `.csv` file or any other file in LazyFrame, you use `scan` instead of `read`. This allows you to read files in lazy mode, whereby your column selections and filtering get pushed down to the scan level. You essentially read only the data necessary for the operations you're performing in your code. You can see that that's much more efficient than reading the whole dataset first and then filtering it down. Again, reading and writing files will be covered in the next chapter.

LazyFrame has similar attributes to DataFrame. However, you'll need to access those via the `.collect_schema()` method. Note that the same method is also available in DataFrame.

> **Note**
>
> Since Polars version 1.0.0, you'll get a performance warning when using LazyFrame attributes such as `.schema`, `.width`, `.dtypes`, and `.columns`. The `.collect_schema()` method replaces those methods. With recent improvements and changes made to the lazy engine, resolving the schema is no longer free and it can be relatively expensive. To solve this, the `.collect_schema()` method was added.

The good news is that it's easy to go back and forth between LazyFrame and DataFrame with `.lazy()` and `.collect()`. This allows you to use LazyFrame where possible and convert to DataFrame if certain operations are not available in the lazy API or if you don't need features such as automatic query optimization and larger-than-RAM processing for your use case.

There's more...

One unique feature of LazyFrame is the ability to inspect the query plan of your code. You can use either the `.show_graph()` or the `.explain()` method. The `.show_graph()` method visualizes the query plan, whereas the `.explain()` method simply prints it out using .show_graph():

```
(
    lf
    .select(pl.col('Name', 'Age'))
    .show_graph()
)
```

The preceding code will return the following output:

> Csv SCAN [../data/titanic_dataset.csv]
> π 2/12;

Figure 1.12 – A query execution plan

π (pi) indicates the column selection and σ (sigma) indicates the filtering conditions.

> **Note**
>
> I haven't introduced the `.filter()` method yet, but just know that it's used to filter data (it's obvious, isn't it?). We'll cover it in a later recipe in this chapter: *Selecting columns and filtering data*.

By default, `.show_graph()` gives you the optimized query plan. You can customize its parameters to choose which optimization to apply. You can find more information on that here: `https://pola-rs.github.io/polars/py-polars/html/reference/lazyframe/api/polars.LazyFrame.show_graph.html`.

For now, here's how to display the non-optimized version:

```
(
    lf
    .select(pl.col('Name', 'Age'))
    .show_graph(optimized=False)
)
```

The preceding code will return the following output:

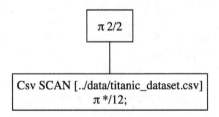

Figure 1.13 – An optimized query execution plan

If you look carefully at both the optimized and the non-optimized version, you'll notice that the former indicates two columns (π **2/12**) whereas the latter indicates all columns (π */**12**).

Let's try calling the `.explain()` method:

```
(
    lf
    .select(pl.col('Name', 'Age'))
    .explain()
)
```

The preceding code will return the following output:

```
'Csv SCAN [../data/titanic_dataset.csv]\nPROJECT 2/12 COLUMNS'
```

Figure 1.14 – A query execution plan in text

You can tweak parameters with the `.explain()` method as well. You can find more information here: `https://pola-rs.github.io/polars/py-polars/html/reference/lazyframe/api/polars.LazyFrame.explain.html`.

The output of the `.explain()` method can be hard to read. To make it more readable, let's try using Python's built-in `print()` function with the separator specified:

```
print(
    lf
    .select(pl.col('Name', 'Age'))
    .explain()
    , sep='\n'
)
```

The preceding code will return the following output:

```
Csv SCAN [../data/titanic_dataset.csv]
PROJECT 2/12 COLUMNS
```

Figure 1.15 – A formatted query execution plan in text

We will dive more into inspecting and optimizing the query plan in *Chapter 12, Testing and Debugging in Polars*

See also

To learn more about LazyFrame, please visit these links:

- `https://pola-rs.github.io/polars/py-polars/html/reference/lazyframe/index.html`
- `https://docs.pola.rs/user-guide/lazy/query-plan/`

Selecting columns and filtering data

In the next few recipes, we'll be covering Polars' essential operations, including column selection, manipulation, and filtering. In this recipe, we'll be covering column selection and filtering specifically.

Selection and filtering are two of the main contexts in which Polars' expressions are evaluated. The power of Polars shines when we utilize expressions in these contexts.

You'll learn how to use some of the most-used DataFrame methods: `.select()`, `.with_columns()`, and `.filter()`.

Getting ready

Read the titanic dataset that we used in the previous recipes if you haven't already:

```
df = pl.read_csv('../data/titanic_dataset.csv')
df.head()
```

How to do it...

We'll first explore selecting columns and then filtering data.

1. Select columns using the `.select()` method. Simply specify one or more column names in the method. Alternatively, you can choose columns with expressions using the `pl.col()` method:

    ```
    df.select(['Survived', 'Ticket', 'Fare']).head()
    ```

    This is what your code will look like when using expressions:

    ```
    df.select(pl.col(['Survived', 'Ticket', 'Fare'])).head()
    ```

    You can also organize the preceding code vertically:

    ```
    df.select(
        pl.col('Survived'),
        pl.col('Ticket'),
        pl.col('Fare')
    ).head()
    ```

The preceding code will return the following output:

shape: (5, 3)

Survived	Ticket	Fare
i64	str	f64
0	"A/5 21171"	7.25
1	"PC 17599"	71.2833
1	"STON/O2. 31012...	7.925
1	"113803"	53.1
0	"373450"	8.05

Figure 1.16 – DataFrame with a few columns

2. Select columns using .with_columns():

```
df.with_columns(['Survived', 'Ticket', 'Fare']).head()
```

Alternatively, you can specify columns explicitly with pl.col():

```
df.with_columns(
    pl.col('Survived'),
    pl.col('Ticket'),
    pl.col('Fare')
).head()
```

The preceding code will return the following output:

shape: (5, 12)

PassengerId	Survived	Pclass	Name	Sex	Age	SibSp	Parch	Ticket	Fare	Cabin	Embarked
i64	i64	i64	str	str	f64	i64	i64	str	f64	str	str
1	0	3	"Braund, Mr. Ow...	"male"	22.0	1	0	"A/5 21171"	7.25	null	"S"
2	1	1	"Cumings, Mrs. ...	"female"	38.0	1	0	"PC 17599"	71.2833	"C85"	"C"
3	1	3	"Heikkinen, Mis...	"female"	26.0	0	0	"STON/O2. 31012...	7.925	null	"S"
4	1	1	"Futrelle, Mrs....	"female"	35.0	1	0	"113803"	53.1	"C123"	"S"
5	0	3	"Allen, Mr. Wil...	"male"	35.0	0	0	"373450"	8.05	null	"S"

Figure 1.17 – Another way to select columns

As a result of the preceding query, all the columns are still selected.

3. Filter data using .filter():

```
df.filter((pl.col('Age') >= 30)).head()
```

The preceding code will return the following output:

shape: (5, 12)

PassengerId	Survived	Pclass	Name	Sex	Age	SibSp	Parch	Ticket	Fare	Cabin	Embarked
i64	i64	i64	str	str	f64	i64	i64	str	f64	str	str
2	1	1	"Cumings, Mrs. ...	"female"	38.0	1	0	"PC 17599"	71.2833	"C85"	"C"
4	1	1	"Futrelle, Mrs....	"female"	35.0	1	0	"113803"	53.1	"C123"	"S"
5	0	3	"Allen, Mr. Wil...	"male"	35.0	0	0	"373450"	8.05	null	"S"
7	0	1	"McCarthy, Mr. ...	"male"	54.0	0	0	"17463"	51.8625	"E46"	"S"
12	1	1	"Bonnell, Miss....	"female"	58.0	0	0	"113783"	26.55	"C103"	"S"

Figure 1.18 – A filtered DataFrame

Let's filter data using multiple conditions:

```
df.filter(
    (pl.col('Age') >= 30) & (pl.col('Sex')=='male')
).head()
```

The preceding code will return the following output:

shape: (5, 12)

PassengerId	Survived	Pclass	Name	Sex	Age	SibSp	Parch	Ticket	Fare	Cabin	Embarked
i64	i64	i64	str	str	f64	i64	i64	str	f64	str	str
5	0	3	"Allen, Mr. Wil...	"male"	35.0	0	0	"373450"	8.05	null	"S"
7	0	1	"McCarthy, Mr. ...	"male"	54.0	0	0	"17463"	51.8625	"E46"	"S"
14	0	3	"Andersson, Mr....	"male"	39.0	1	5	"347082"	31.275	null	"S"
21	0	2	"Fynney, Mr. Jo...	"male"	35.0	0	0	"239865"	26.0	null	"S"
22	1	2	"Beesley, Mr. L...	"male"	34.0	0	0	"248698"	13.0	"D56"	"S"

Figure 1.19 – Multiple filtering conditions

How it works...

Both the `.select()` and `.with_columns()` methods are used for column selection and manipulation. Notice that the output between the `.select()` and `.with_columns()` methods is different, even though the syntax is very similar in the preceding examples.

The difference between the `.select()` and `.with_columns()` methods is that `.select()` drops the columns that are not selected, whereas `.with_columns()` replaces existing columns with the same name. When you only specify existing columns inside `.with_columns()`, you're basically selecting all columns.

The `.filter()` method simply filters data based on the condition(s) that you write with expressions. You'd need to use & or | for and and or operators.

There's more...

In Polars, you can select columns like you can do in pandas:

```
df[['Age', 'Sex']].head()
```

The preceding code will return the following output:

shape: (5, 2)

Age	Sex
f64	str
22.0	"male"
38.0	"female"
26.0	"female"
35.0	"female"
35.0	"male"

Figure 1.20 – pandas's way of selecting columns

> **Note**
>
> The fact that you can do something doesn't mean that you should. The best practice is to utilize expressions as much as possible. Expressions help you use Polars to its full potential, including using parallel execution and query optimizations.

When you start using expressions, your code will become more concise and readable with the use of method chaining. We'll cover method chaining later in a recipe called *Understanding method chaining*.

It's worth introducing a few more advanced, convenient ways of selecting columns in this section.

One of them is selecting columns by **regular expressions** (**regex**). This example selects columns whose character length is less than or equal to 4:

```
df.select(pl.col('^[a-zA-Z]{0,4}$')).head()
```

The preceding code will return the following output:

shape: (5, 4)

Name	Sex	Age	Fare
str	str	f64	f64
"Braund, Mr. Ow...	"male"	22.0	7.25
"Cumings, Mrs. ...	"female"	38.0	71.2833
"Heikkinen, Mis...	"female"	26.0	7.925
"Futrelle, Mrs....	"female"	35.0	53.1
"Allen, Mr. Wil...	"male"	35.0	8.05

Figure 1.21 – Selecting columns with regex

As a side note, the following website is useful when using regex: `https://regexr.com`.

Another way of selecting columns is by using data types. Let's select columns whose data type is `string`:

```
df.select(pl.col(pl.String)).head()
```

The preceding code will return the following output:

shape: (5, 5)

Name	Sex	Ticket	Cabin	Embarked
str	str	str	str	str
"Braund, Mr. Ow...	"male"	"A/5 21171"	null	"S"
"Cumings, Mrs. ...	"female"	"PC 17599"	"C85"	"C"
"Heikkinen, Mis...	"female"	"STON/O2. 31012...	null	"S"
"Futrelle, Mrs....	"female"	"113803"	"C123"	"S"
"Allen, Mr. Wil...	"male"	"373450"	null	"S"

Figure 1.22 – Column selection with data types

A more advanced way of selecting columns is by using functions available in the `selectors` namespace. Here's a simple example:

```
import polars.selectors as cs
df.select(cs.numeric()).head()
```

The preceding code will return the following output:

shape: (5, 5)

Name	Sex	Ticket	Cabin	Embarked
str	str	str	str	str
"Braund, Mr. Ow...	"male"	"A/5 21171"	null	"S"
"Cumings, Mrs. ...	"female"	"PC 17599"	"C85"	"C"
"Heikkinen, Mis...	"female"	"STON/O2. 31012...	null	"S"
"Futrelle, Mrs....	"female"	"113803"	"C123"	"S"
"Allen, Mr. Wil...	"male"	"373450"	null	"S"

Figure 1.23 – Column selection with selectors

Here's how to use the `cs.matches()` function, selecting columns that include words "se" or "ed":

```
df.select(cs.matches('se|ed')).head()
```

The preceding code will return the following output:

shape: (5, 3)

PassengerId	Survived	Embarked
i64	i64	str
1	0	"S"
2	1	"C"
3	1	"S"
4	1	"S"
5	0	"S"

Figure 1.24 – Another way to select columns with selectors

There is a lot more you can do with selectors such as **setting operations** (e.g., union or intersection). For additional information about which selectors functions are available, refer to this Polars documentation: https://pola-rs.github.io/polars/py-polars/html/reference/selectors.html.

See also

Please refer to these pages in the Polars documentation for additional information:

- https://pola-rs.github.io/polars/user-guide/expressions/column-selections

- https://pola-rs.github.io/polars/py-polars/html/reference/dataframe/api/polars.DataFrame.filter.html

Creating, modifying, and deleting columns

The key methods we'll cover in this recipe are .select(), .with_columns(), and .drop(). We've seen in the previous recipe that both .select() and .with_columns() are essential for column selection in Polars.

In this recipe, you'll learn how to leverage those methods to create, modify, and delete columns using Polars' expressions.

Getting ready

This recipe requires the titanic dataset. Read it into your code by typing the following:

```
df = pl.read_csv('../data/titanic_dataset.csv')
```

How to do it...

Let's dive into the recipe. Here are the steps:

1. Create a column based on another column:

```
df.with_columns(
    pl.col('Fare').max().alias('Max Fare')
).head()
```

The preceding code will return the following output:

shape: (5, 13)

PassengerId	Survived	Pclass	Name	Sex	Age	SibSp	Parch	Ticket	Fare	Cabin	Embarked	Max Fare
i64	i64	i64	str	str	f64	i64	i64	str	f64	str	str	f64
1	0	3	"Braund, Mr. Ow...	"male"	22.0	1	0	"A/5 21171"	7.25	null	"S"	512.3292
2	1	1	"Cumings, Mrs. ...	"female"	38.0	1	0	"PC 17599"	71.2833	"C85"	"C"	512.3292
3	1	3	"Heikkinen, Mis...	"female"	26.0	0	0	"STON/O2. 31012...	7.925	null	"S"	512.3292
4	1	1	"Futrelle, Mrs....	"female"	35.0	1	0	"113803"	53.1	"C123"	"S"	512.3292
5	0	3	"Allen, Mr. Wil...	"male"	35.0	0	0	"373450"	8.05	null	"S"	512.3292

Figure 1.25 – A DataFrame with a new column

We added a new column called `max_fare`. Its value is the max of the `Fare` column. We'll cover aggregations in more detail in a later chapter.

You can name your column without using `.alias()`. You'll need to specify the name at the beginning of your expression. Note that you won't be able to use spaces in the column name with this approach:

```
df.with_columns(
    max_fare=pl.col('Fare').max()
).head()
```

The preceding code will return the following output:

shape: (5, 13)

PassengerId	Survived	Pclass	Name	Sex	Age	SibSp	Parch	Ticket	Fare	Cabin	Embarked	max_fare
i64	i64	i64	str	str	f64	i64	i64	str	f64	str	str	f64
1	0	3	"Braund, Mr. Ow...	"male"	22.0	1	0	"A/5 21171"	7.25	null	"S"	512.3292
2	1	1	"Cumings, Mrs. ...	"female"	38.0	1	0	"PC 17599"	71.2833	"C85"	"C"	512.3292
3	1	3	"Heikkinen, Mis...	"female"	26.0	0	0	"STON/O2. 31012...	7.925	null	"S"	512.3292
4	1	1	"Futrelle, Mrs....	"female"	35.0	1	0	"113803"	53.1	"C123"	"S"	512.3292
5	0	3	"Allen, Mr. Wil...	"male"	35.0	0	0	"373450"	8.05	null	"S"	512.3292

Figure 1.26 – A different way to name a new column

If you don't specify a new column name, then the base column will be overwritten:

```
df.with_columns(
    pl.col('Fare').max()
).head()
```

The preceding code will return the following output:

shape: (5, 12)

PassengerId	Survived	Pclass	Name	Sex	Age	SibSp	Parch	Ticket	Fare	Cabin	Embarked
i64	i64	i64	str	str	f64	i64	i64	str	f64	str	str
1	0	3	"Braund, Mr. Ow…	"male"	22.0	1	0	"A/5 21171"	512.3292	null	"S"
2	1	1	"Cumings, Mrs. …	"female"	38.0	1	0	"PC 17599"	512.3292	"C85"	"C"
3	1	3	"Heikkinen, Mis…	"female"	26.0	0	0	"STON/O2. 31012…	512.3292	null	"S"
4	1	1	"Futrelle, Mrs.…	"female"	35.0	1	0	"113803"	512.3292	"C123"	"S"
5	0	3	"Allen, Mr. Wil…	"male"	35.0	0	0	"373450"	512.3292	null	"S"

Figure 1.27 – A new column with the same name as the base column

To demonstrate how you can use multiple expressions for a column, let's add another logic to this column:

```
df.with_columns(
    (pl.col('Fare').max() - pl.col('Fare').mean()).alias('Max Fare -
Avg Fare')
).head()
```

The preceding code will return the following output:

shape: (5, 13)

PassengerId	Survived	Pclass	Name	Sex	Age	SibSp	Parch	Ticket	Fare	Cabin	Embarked	Max Fare - Avg Fare
i64	i64	i64	str	str	f64	i64	i64	str	f64	str	str	f64
1	0	3	"Braund, Mr. Ow…	"male"	22.0	1	0	"A/5 21171"	7.25	null	"S"	480.124992
2	1	1	"Cumings, Mrs. …	"female"	38.0	1	0	"PC 17599"	71.2833	"C85"	"C"	480.124992
3	1	3	"Heikkinen, Mis…	"female"	26.0	0	0	"STON/O2. 31012…	7.925	null	"S"	480.124992
4	1	1	"Futrelle, Mrs.…	"female"	35.0	1	0	"113803"	53.1	"C123"	"S"	480.124992
5	0	3	"Allen, Mr. Wil…	"male"	35.0	0	0	"373450"	8.05	null	"S"	480.124992

Figure 1.28 – A new column with more complex expressions

We added a column that calculates the max and mean of the Fare column and does a subtraction. This is just one example of how you can use Polars' expressions.

1. Create a column with a literal value using the pl.lit() method:

    ```
    df.with_columns(pl.lit('Titanic')).head()
    ```

    The preceding code will return the following output:

shape: (5, 13)

PassengerId	Survived	Pclass	Name	Sex	Age	SibSp	Parch	Ticket	Fare	Cabin	Embarked	literal
i64	i64	i64	str	str	f64	i64	i64	str	f64	str	str	str
1	0	3	"Braund, Mr. Ow…	"male"	22.0	1	0	"A/5 21171"	7.25	null	"S"	"Titanic"
2	1	1	"Cumings, Mrs. …	"female"	38.0	1	0	"PC 17599"	71.2833	"C85"	"C"	"Titanic"
3	1	3	"Heikkinen, Mis…	"female"	26.0	0	0	"STON/O2. 31012…	7.925	null	"S"	"Titanic"
4	1	1	"Futrelle, Mrs.…	"female"	35.0	1	0	"113803"	53.1	"C123"	"S"	"Titanic"
5	0	3	"Allen, Mr. Wil…	"male"	35.0	0	0	"373450"	8.05	null	"S"	"Titanic"

Figure 1.29 – The output with literal values

2. Add a row count with `.with_row_index()`:

```
df.with_row_index().head()
```

The preceding code will return the following output:

shape: (5, 13)

index	PassengerId	Survived	Pclass	Name	Sex	Age	SibSp	Parch	Ticket	Fare	Cabin	Embarked
u32	i64	i64	i64	str	str	f64	i64	i64	str	f64	str	str
0	1	0	3	"Braund, Mr. Owen Harris"	"male"	22.0	1	0	"A/5 21171"	7.25	null	"S"
1	2	1	1	"Cumings, Mrs. John Bradley (Fl...	"female"	38.0	1	0	"PC 17599"	71.2833	"C85"	"C"
2	3	1	3	"Heikkinen, Miss. Laina"	"female"	26.0	0	0	"STON/O2. 3101282"	7.925	null	"S"
3	4	1	1	"Futrelle, Mrs. Jacques Heath (...	"female"	35.0	1	0	"113803"	53.1	"C123"	"S"
4	5	0	3	"Allen, Mr. William Henry"	"male"	35.0	0	0	"373450"	8.05	null	"S"

Figure 1.30 – The output with a row number

3. Modify values in a column:

```
df.with_columns(pl.col('Sex').str.to_titlecase()).head()
```

The preceding code will return the following output:

shape: (5, 12)

PassengerId	Survived	Pclass	Name	Sex	Age	SibSp	Parch	Ticket	Fare	Cabin	Embarked
i64	i64	i64	str	str	f64	i64	i64	str	f64	str	str
1	0	3	"Braund, Mr. Ow...	"Male"	22.0	1	0	"A/5 21171"	7.25	null	"S"
2	1	1	"Cumings, Mrs. ...	"Female"	38.0	1	0	"PC 17599"	71.2833	"C85"	"C"
3	1	3	"Heikkinen, Mis...	"Female"	26.0	0	0	"STON/O2. 31012...	7.925	null	"S"
4	1	1	"Futrelle, Mrs....	"Female"	35.0	1	0	"113803"	53.1	"C123"	"S"
5	0	3	"Allen, Mr. Wil...	"Male"	35.0	0	0	"373450"	8.05	null	"S"

Figure 1.31 – The output of the modified column

We transformed the Sex column into title case. `.str` is what gives you access to string methods in Polars, which we'll cover in *Chapter 6, Performing String Manipulations*.

1. You can delete a column with the help of the following code:

```
df.drop(['Pclass', 'Name', 'SibSp', 'Parch', 'Ticket', 'Cabin',
'Embarked']).head()
```

The preceding code will return the following output:

shape: (5, 5)

PassengerId	Survived	Sex	Age	Fare
i64	i64	str	f64	f64
1	0	"male"	22.0	7.25
2	1	"female"	38.0	71.2833
3	1	"female"	26.0	7.925
4	1	"female"	35.0	53.1
5	0	"male"	35.0	8.05

Figure 1.32 – The output after dropping columns

2. You can use .select() instead to choose the columns that you want to keep:

```
df.select(['PassengerId', 'Survived', 'Sex', 'Age', 'Fare']).
head()
```

The preceding code will return the following output:

shape: (5, 5)

PassengerId	Survived	Sex	Age	Fare
i64	i64	str	f64	f64
1	0	"male"	22.0	7.25
2	1	"female"	38.0	71.2833
3	1	"female"	26.0	7.925
4	1	"female"	35.0	53.1
5	0	"male"	35.0	8.05

Figure 1.33 – DataFrame with selected columns

How it works...

The pl.lit() method can be used whenever you want to specify a literal or constant value. You can use not only a string value but also various data types such as integer, boolean, list, and so on.

When creating or adding a new column, there are three ways you can name it:

- Use .alias().

- Define the column name at the beginning of your expression, like the one you saw earlier: max_fare=pl.col('Fare').max(). You can't use spaces in your column name.

- Don't specify the column name, which would replace the existing column if the new column were created based on another column. Alternatively, the column will be named literal when using pl.lit().

Both the `.select()` and `.with_columns()` methods can create and modify columns. The difference is in whether you keep the unspecified columns or drop them. Essentially, you can use the `.select()` method for dropping columns while adding new columns. That way, you may avoid using both the `.with_columns()` and `.drop()` methods in combination when `.select()` alone can do the job.

Also, note that new or modified columns don't persist when using the `.select()` or `.with_columns()` methods. You'll need to store the result into a variable if needed:

```
df = df.with_columns(
    pl.col('Fare').max()
)
```

There's more...

For best practice, you should put all your expressions into one method where possible instead of using multiple `.with_columns()`, for example. This makes sure that expressions are executed in parallel, whereas if you use multiple `.with_columns()`, then Polars's engine might not recognize that they run in parallel.

You should write your code like this:

```
best_practice = (
    df.with_columns(
        pl.col('Fare').max().alias('Max Fare'),
        pl.lit('Titanic'),
        pl.col('Sex').str.to_titlecase()
    )
)
```

Avoid writing your code like this:

```
not_so_good_practice = (
    df
    .with_columns(pl.col('Fare').max().alias('Max Fare'))
    .with_columns(pl.lit('Titanic'))
    .with_columns(pl.col('Sex').str.to_titlecase())
)
```

Both of the preceding queries produce the following output:

shape: (5, 14)

PassengerId	Survived	Pclass	Name	Sex	Age	SibSp	Parch	Ticket	Fare	Cabin	Embarked	Max Fare	literal
i64	i64	i64	str	str	f64	i64	i64	str	f64	str	str	f64	str
1	0	3	"Braund, Mr. Ow...	"Male"	22.0	1	0	"A/5 21171"	7.25	null	"S"	512.3292	"Titanic"
2	1	1	"Cumings, Mrs. ...	"Female"	38.0	1	0	"PC 17599"	71.2833	"C85"	"C"	512.3292	"Titanic"
3	1	3	"Heikkinen, Mis...	"Female"	26.0	0	0	"STON/O2. 31012...	7.925	null	"S"	512.3292	"Titanic"
4	1	1	"Futrelle, Mrs....	"Female"	35.0	1	0	"113803"	53.1	"C123"	"S"	512.3292	"Titanic"
5	0	3	"Allen, Mr. Wil...	"Male"	35.0	0	0	"373450"	8.05	null	"S"	512.3292	"Titanic"

Figure 1.34 – The output with new columns added

> **Note**
>
> You won't be able to add a new column on top of another new column you're trying to define in the same method (such as the .with_columns() method). The only time when you'll need to use multiple methods is when your new column depends on another new column in your dataset that doesn't yet exist.

See also

Please refer to these resources for more information:

- https://pola-rs.github.io/polars/user-guide/concepts/expressions/

- https://www.pola.rs/posts/the-expressions-api-in-polars-is-amazing/

- https://pola-rs.github.io/polars/user-guide/concepts/contexts/

- https://pola-rs.github.io/polars/py-polars/html/reference/expressions/api/polars.lit.html

- https://pola-rs.github.io/polars/py-polars/html/reference/expressions/api/polars.Expr.alias.html

Understanding method chaining

Method chaining is a technique or way of structuring your code. It's commonly used across DataFrame libraries such as pandas and PySpark. As the name tells you, it means that you chain methods one after another. This makes your code more readable, concise, and maintainable. It follows a natural flow from one operation to another, which makes your code easy to follow. All of that helps you focus on the data transformation logic and problems you're trying to solve.

The good news is that Polars is a good fit for method chaining. Polars utilizes expressions and other methods that can easily be stacked on each other.

Getting ready

This recipe requires the titanic dataset. Make sure to read it into a DataFrame:

```
df = pl.read_csv('../data/titanic_dataset.csv')
```

How to do it...

Let's say that you're doing a few operations on the dataset. First, we will predefine the columns that we want to select:

```
cols = ['Name', 'Sex', 'Age', 'Fare', 'Cabin', 'Pclass', 'Survived']
```

If you're not using method chaining, you might want to write code like this:

```
df = df.select(cols)
df = df.filter(pl.col('Age')>=35)
df = df.sort(by=['Age', 'Name'])
```

When you use method chaining, it'd look like this:

```
df = df.select(cols).filter(pl.col('Age')>=35).sort(by=['Age',
'Name'])
```

To go one step further, let's stack these methods vertically. This is the preferred way of writing your code with method chaining:

```
df = (
    df
    .select(cols)
    .filter(pl.col('Age')>=35)
    .sort(by=['Age', 'Name'])
)
```

All of the preceding code produces the same output:

shape: (5, 7)

Name	Sex	Age	Fare	Cabin	Pclass	Survived
str	str	f64	f64	str	i64	i64
"Abbott, Mrs. S...	"female"	35.0	20.25	null	3	1
"Allen, Mr. Wil...	"male"	35.0	8.05	null	3	0
"Asim, Mr. Adol...	"male"	35.0	7.05	null	3	0
"Bissette, Miss...	"female"	35.0	135.6333	"C99"	1	1
"Brocklebank, M...	"male"	35.0	8.05	null	3	0

Figure 1.35 – The output after column selection, filtering, and sorting

How it works...

The first example I showed defines each method line by line, storing each result in a variable each time. The last example involved method chaining, aligning the beginning of each method vertically. Some users don't even know that you can stack your methods on top of each other, especially users who are just getting started. You might have a habit of defining your transformations line by line, like in the first example.

Having looked at a few examples, which pattern do you think is best? I'd say the one using method chaining, stacking each method vertically. Aligning the beginning of each method helps with readability. Having all the logic in the same place makes it easier to maintain the code and figure things out later. It also helps you streamline your workflows by making your code more concise and ensuring that it is organized in a logical way.

How does this help with testing and debugging though? You can comment out or add another method within the parentheses to test the result:

```
df = (
    df
    .select(cols)
    # .filter(pl.col('Age')>=35)
    .sort(by=['Age', 'Name'])
)
df.head()
```

The preceding code will return the following output:

shape: (5, 7)

Name	Sex	Age	Fare	Cabin	Pclass	Survived
str	str	f64	f64	str	i64	i64
"Abbott, Mrs. S...	"female"	35.0	20.25	null	3	1
"Allen, Mr. Wil...	"male"	35.0	8.05	null	3	0
"Asim, Mr. Adol...	"male"	35.0	7.05	null	3	0
"Bissette, Miss...	"female"	35.0	135.6333	"C99"	1	1
"Brocklebank, M...	"male"	35.0	8.05	null	3	0

Figure 1.36 – The first five rows without the filtering condition

We'll cover testing and debugging in more detail in *Chapter 12, Testing and Debugging in Polars*.

One caveat is that when your chain is too long, it may make your code hard to read and work with. This increased complexity that comes with a long chain can make your debugging hard, too. It can become challenging to understand each intermediary step in a long chain. In that case, you should break your logic down into smaller pieces to help reduce the complexity and length of your chain. With all of that said, it all comes down to the fact that a balance is needed to make testing your code feasible.

In the interest of full disclosure, remember that you don't have an obligation to use method chaining. If it feels more comfortable or appropriate to write your code line by line separately, that's all good and fine. Method chaining is just another practice, and many people find it helpful. I can confidently say that method chaining has done me more good than harm.

There's more...

When you stack your methods vertically, you can also use backslashes instead of using parentheses:

```
df = df \
    .select(cols) \
    .filter(pl.col('Age')>=35) \
    .sort(by=['Age', 'Name'])
```

I have to say that adding a backslash for each method is a little bit of work. Also, if you comment out the last method in the chain for testing and debugging purposes, it messes up the whole chain because you can't end your code with a backslash. I'd choose using parentheses over backslashes any day.

See also

These are useful resources to learn more about method chaining:

- `https://ponder.io/professional-pandas-the-pandas-assign-method-and-chaining/`

- `https://tomaugspurger.net/posts/method-chaining/`

Processing larger-than-RAM datasets

One of the outstanding features of Polars is its streaming mode. It's part of the lazy API and it allows you to process data that is larger than the memory available on your machine. With streaming mode, you let your machine handle huge data by processing it in batches. You would not be able to process such large data otherwise.

One thing to keep in mind is that not all lazy operations are supported in streaming mode, as it's still in development. You can still use any lazy operation in your query, but ultimately, the Polars engine will determine whether the operation can be executed in streaming or not. If the answer is no, then Polars runs the query using non-streaming mode. We can expect that this feature will include more lazy operations and become more sophisticated over time.

In this recipe, we'll demonstrate how streaming mode works by creating a simple query to read a .csv file that's larger than the available RAM on a machine and process it using streaming mode.

Getting ready

You'd need a dataset that's larger than the available RAM on your machine to test streaming mode. I'm using a taxi trips dataset, which has over 80 GB on disk. You can download the dataset from this website: `https://data.cityofchicago.org/Transportation/Taxi-Trips-2013-2023-/wrvz-psew/about_data`.

How to do it...

Here are the steps for the recipe.

1. Import the Polars library:

```
import polars as pl
```

2. Read the csv file in streaming mode by adding a `streaming=True` parameter inside `.collect()`. The file name string should specify where your file is located (mine is in my Downloads folder):

```
taxi_trips = (
    pl.scan_csv('~/Downloads/Taxi_Trips.csv')
    .collect(streaming=True)
)
```

3. Check the first five rows with `.head()` to see what the data looks like:

```
taxi_trips.head()
```

The preceding code will return the following output:

shape: (5, 23)

Trip ID	Taxi ID	Trip Start Timestamp	Trip End Timestamp	Trip Seconds	Trip Miles	Pickup Census Tract	Dropoff Census Tract	Pickup Community Area	Dropoff Community Area	Fare	Tips	Tolls	Extras	Trip Total	Payment Type	Company	
str	str	str	str	i64	f64	i64	i64	i64	i64	f64	f64	f64	f64	f64	str	str	
"2c4c7c96c03236...	"f68fe1183e6d67...	"10/23/2016 10:...	"10/23/2016 10:...	360	1.1	17031320100	17031320100	32	32	6.75	0.0	0.0	1.5	8.25	"Cash"	"Medallion Leas...	41
"d3a30f8612e040...	"97ca0053c70788...	"10/23/2016 07:...	"10/23/2016 07:...	300	1.0	17031320100	17031081700	32	8	6.25	0.0	0.0	0.0	6.25	"Cash"	"Medallion Leas...	41
"95178b9e52c980...	"f0b96329a7e390...	"10/27/2016 08:...	"10/27/2016 09:...	636	2.23	17031833100	17031081403	28	8	9.75	0.0	0.0	0.0	9.75	"Cash"	"City Service"	4
"b221f753f4fce4...	"11b7c42b4edd22...	"10/27/2016 08:...	"10/27/2016 08:...	1000	2.13	17031320600	17031330100	32	33	10.75	0.0	0.0	1.5	12.25	"Cash"	"Chicago Carria...	4
"d6d73a306f0e46...	"d4949520604804d4...	"10/27/2016 05:...	"10/27/2016 05:...	540	1.6	17031081900	17031833000	8	28	7.5	0.0	0.0	0.0	7.5	"Cash"	"Taxi Affiliati...	41

Figure 1.37 – The first five rows of the taxi trip dataset

How it works...

There are two things you should be aware of in the example code:

- It uses `.scan_read()` instead of `.read_csv()`
- A parameter is specified in `.collect()`. It becomes `.collect(streaming=True)`.

We will enable streaming mode by setting `streaming=True` inside the `.collect()` method. In this specific example, I'm only reading a `.csv` file, nothing complex. I'm using the `.scan_read()` method to read with lazy mode.

In theory, without streaming mode, I wouldn't be able to process this dataset. This is because my laptop has 64 GB of RAM (yes, my laptop has a decent amount of memory!), which is lower than the size of the dataset on disk, which is more than 80 GB.

It took about two minutes for my laptop to process the data in streaming mode. Without streaming mode, I would get an out-of-memory error. You can confirm this by running your code without `streaming=True` in the `.collect()` method.

There's more...

If you're doing other operations other than reading the data, such as aggregations and filtering, then Polars (with LazyFrame) might be able to optimize your query so that it doesn't need to read the whole dataset in memory. This means that you might not even need to utilize streaming mode to work with data larger than your RAM. Aggregations and filtering essentially summarize the data or reduce the number of rows, which leads to not needing to read in the whole dataset.

Let's say that you apply a simple group by and aggregation over a column like the one in the following code. You'll see that you can run it without using streaming mode (depending on your chosen dataset and the available RAM on your machine):

```
trip_total_by_pay_type = (
    pl.scan_csv('~/Downloads/Taxi_Trips.csv')
    .group_by('Payment Type')
    .agg(pl.col('Trip Total').sum())
    .collect()
)
trip_total_by_pay_type.head()
```

The preceding code will return the following output:

shape: (5, 2)

Payment Type	Trip Total
str	f64
"Dispute"	1.3884e6
"Split"	64668.43
"Prepaid"	40650.03
"Credit Card"	1.7685e9
"Prcard"	4.0122e7

Figure 1.38 – Trip total by payment type

With that said, it may still be a good idea to use `streaming=True` when there is a possibility that the size of the dataset goes over your available RAM or that data may grow in size over time.

See also

Please refer to the streaming API page in Polars's documentation: `https://pola-rs.github.io/polars-book/user-guide/concepts/streaming/`.

2
Reading and Writing Files

Reading and writing files is a fundamental step in your data workflows. Your data processing pipelines have both input and output. Learning and knowing how to input and output your data effectively is an essential component for your successful Polars implementations.

Polars provides a way to read and write files while in lazy mode by utilizing the `scan` and `sink` methods. They help push down predicates (filters) and projections (column selections) to the scan level and write the output that is larger than RAM streaming results to disk..

In this chapter, you'll be learning how to read and write various types of file formats.

We will explore the following recipes in this chapter:

- Reading and writing CSV files
- Reading and writing Parquet files
- Reading and writing Delta Lake tables
- Reading and writing JSON files
- Reading and writing Excel files
- Reading and writing other data file formats
- Reading and writing multiple files
- Working with databases

Technical requirements

You can download the datasets and code for this chapter from the GitHub repository:

- You can find the data here: `https://github.com/PacktPublishing/Polars-Cookbook/tree/main/data`
- The code is located here: `https://github.com/PacktPublishing/Polars-Cookbook/tree/main/Chapter02`

Also, it is assumed that you have installed the Polars library in your Python environment:

```
>>> pip install polars
```

It is also assumed that you imported it in your code:

```
import polars as pl
```

Reading and writing CSV files

Comma-separated values (**CSV**) is one of the most commonly used file formats for storing data. The structure or the way in which you read a CSV file may be familiar to you if you have worked with another DataFrame library such as pandas.

In this recipe, we'll examine how to read and write a CSV file in Polars with some parameters. We'll also look at how we can do the same in a LazyFrame.

How to do it...

Here are the steps and examples for how to read and write CSV files in Polars:

1. Read the `customer_shopping_data.csv` dataset into a DataFrame:

    ```
    df = pl.read_csv('../data/customer_shopping_data.csv')
    df.head()
    ```

    The preceding code will return the following output:

    shape: (5, 10)

invoice_no	customer_id	gender	age	category	quantity	price	payment_method	invoice_date	shopping_mall
str	str	str	i64	str	i64	f64	str	str	str
"I138884"	"C241288"	"Female"	28	"Clothing"	5	1500.4	"Credit Card"	"5/8/2022"	"Kanyon"
"I317333"	"C111565"	"Male"	21	"Shoes"	3	1800.51	"Debit Card"	"12/12/2021"	"Forum Istanbul...
"I127801"	"C266599"	"Male"	20	"Clothing"	1	300.08	"Cash"	"9/11/2021"	"Metrocity"
"I173702"	"C988172"	"Female"	66	"Shoes"	5	3000.85	"Credit Card"	"16/05/2021"	"Metropol AVM"
"I337046"	"C189076"	"Female"	53	"Books"	4	60.6	"Cash"	"24/10/2021"	"Kanyon"

    Figure 2.1 – The first five rows of the customer shopping dataset

2. If the CSV file doesn't have a header, Polars would treat the first row as the header:

    ```
    df = pl.read_csv('../data/customer_shopping_data_no_header.csv')
    df.head()
    ```

The preceding code will return the following output:

shape: (5, 10)

I138884	C241288	Female	28	Clothing	5	1500.4	Credit Card	5/8/2022	Kanyon
str	str	str	i64	str	i64	f64	str	str	str
"I317333"	"C111565"	"Male"	21	"Shoes"	3	1800.51	"Debit Card"	"12/12/2021"	"Forum Istanbul...
"I127801"	"C266599"	"Male"	20	"Clothing"	1	300.08	"Cash"	"9/11/2021"	"Metrocity"
"I173702"	"C988172"	"Female"	66	"Shoes"	5	3000.85	"Credit Card"	"16/05/2021"	"Metropol AVM"
"I337046"	"C189076"	"Female"	53	"Books"	4	60.6	"Cash"	"24/10/2021"	"Kanyon"
"I227836"	"C657758"	"Female"	28	"Clothing"	5	1500.4	"Credit Card"	"24/05/2022"	"Forum Istanbul...

Figure 2.2 – The first row is set as the header

3. You can read the same file with a parameter specified to fix the header:

```
df = pl.read_csv('../data/customer_shopping_data_no_header.csv',
has_header=False)
df.head()
```

The preceding code will return the following output:

shape: (5, 10)

column_1	column_2	column_3	column_4	column_5	column_6	column_7	column_8	column_9	column_10
str	str	str	i64	str	i64	f64	str	str	str
"I138884"	"C241288"	"Female"	28	"Clothing"	5	1500.4	"Credit Card"	"5/8/2022"	"Kanyon"
"I317333"	"C111565"	"Male"	21	"Shoes"	3	1800.51	"Debit Card"	"12/12/2021"	"Forum Istanbul...
"I127801"	"C266599"	"Male"	20	"Clothing"	1	300.08	"Cash"	"9/11/2021"	"Metrocity"
"I173702"	"C988172"	"Female"	66	"Shoes"	5	3000.85	"Credit Card"	"16/05/2021"	"Metropol AVM"
"I337046"	"C189076"	"Female"	53	"Books"	4	60.6	"Cash"	"24/10/2021"	"Kanyon"

Figure 2.3 – The column names are auto-assigned

4. Add a parameter to specify the column names:

```
column_names = ['invoice_no', 'customer_id', 'gender',
                'age', 'category', 'quantity',
                'price', 'payment_method',
                'invoice_date', 'shopping_mall']
df = pl.read_csv('../data/customer_shopping_data_no_header.csv',
        has_header=False,
        new_columns=column_names)
df.head()
```

The preceding code will return the following output:

shape: (5, 10)

invoice_no	customer_id	gender	age	category	quantity	price	payment_method	invoice_date	shopping_mall
str	str	str	i64	str	i64	f64	str	str	str
"I138884"	"C241288"	"Female"	28	"Clothing"	5	1500.4	"Credit Card"	"5/8/2022"	"Kanyon"
"I317333"	"C111565"	"Male"	21	"Shoes"	3	1800.51	"Debit Card"	"12/12/2021"	"Forum Istanbul…
"I127801"	"C266599"	"Male"	20	"Clothing"	1	300.08	"Cash"	"9/11/2021"	"Metrocity"
"I173702"	"C988172"	"Female"	66	"Shoes"	5	3000.85	"Credit Card"	"16/05/2021"	"Metropol AVM"
"I337046"	"C189076"	"Female"	53	"Books"	4	60.6	"Cash"	"24/10/2021"	"Kanyon"

Figure 2.4 – Specified column names upon reading

5. Make sure that the dates are parsed upon reading:

```
df = pl.read_csv('../data/customer_shopping_data_no_header.csv',
    has_header=False,
    new_columns=column_names,
    try_parse_dates=True)
df.head()
```

The preceding code will return the following output:

shape: (5, 10)

invoice_no	customer_id	gender	age	category	quantity	price	payment_method	invoice_date	shopping_mall
str	str	str	i64	str	i64	f64	str	date	str
"I138884"	"C241288"	"Female"	28	"Clothing"	5	1500.4	"Credit Card"	2022-08-05	"Kanyon"
"I317333"	"C111565"	"Male"	21	"Shoes"	3	1800.51	"Debit Card"	2021-12-12	"Forum Istanbul…
"I127801"	"C266599"	"Male"	20	"Clothing"	1	300.08	"Cash"	2021-11-09	"Metrocity"
"I173702"	"C988172"	"Female"	66	"Shoes"	5	3000.85	"Credit Card"	2021-05-16	"Metropol AVM"
"I337046"	"C189076"	"Female"	53	"Books"	4	60.6	"Cash"	2021-10-24	"Kanyon"

Figure 2.5 – Parsed dates upon reading the file

6. You can also specify data types:

```
df = pl.read_csv('../data/customer_shopping_data_no_header.csv',
    has_header=False,
    new_columns=column_names,
    try_parse_dates=True,
    schema_overrides={'age': pl.Int8, 'quantity':
pl.Int32})
df.head()
```

The preceding code will return the following output:

shape: (5, 10)

invoice_no	customer_id	gender	age	category	quantity	price	payment_method	invoice_date	shopping_mall
str	str	str	i8	str	i32	f64	str	date	str
"I138884"	"C241288"	"Female"	28	"Clothing"	5	1500.4	"Credit Card"	2022-08-05	"Kanyon"
"I317333"	"C111565"	"Male"	21	"Shoes"	3	1800.51	"Debit Card"	2021-12-12	"Forum Istanbul...
"I127801"	"C266599"	"Male"	20	"Clothing"	1	300.08	"Cash"	2021-11-09	"Metrocity"
"I173702"	"C988172"	"Female"	66	"Shoes"	5	3000.85	"Credit Card"	2021-05-16	"Metropol AVM"
"I337046"	"C189076"	"Female"	53	"Books"	4	60.6	"Cash"	2021-10-24	"Kanyon"

Figure 2.6 – Specified data types upon reading

7. Now, let's write a DataFrame to a CSV file with a few parameters specified:

```
df.write_csv('../data/output/shopping_data_output.csv',
             include_header=False,
             separator=',')
```

How it works...

has_header accepts a CSV file without the header. Polars auto-populates column names if the file doesn't contain the header. You can also utilize similar parameters such as skip_rows and separator to adjust the input file format upon reading.

With the new_columns parameter, you can replace the existing column names in the order in which your new column names are defined in your code.

What the try_parse_dates parameter does is try to auto-parse date columns. If the date column cannot be parsed, it appears as the pl.Utf8 type. The use_pyarrow parameter also tries to parse dates automatically, but it uses pyarrow's CSV parser.

With the schema_overrides parameter, you can overwrite data types for specific columns. The schema parameter works similarly to the schema_overrides parameter. The difference is that schema expects the complete schema.

There are many other parameters that you can use when reading CSV files. There are not only ones that have to do with the structure of the input file but also ones that focus on performance and how Polars should handle the input file. Parameters such as n_threads, batch_size, and low_memory might come in handy if you want to have more control over performance and resource usage. The storage_options parameter lets you work with cloud storage, which is a topic that we'll cover in *Chapter 11, Working with Common Cloud Data Sources*.

Also, you can read a CSV file in batches, extracting a DataFrame per batch. Since that topic falls outside of the scope of this book, you can refer to the Polars documentation for more information at https://pola-rs.github.io/polars/py-polars/html/reference/api/polars.read_csv_batched.html.

There are fewer parameters available for writing a CSV file. You can specify the output file format with parameters such as `has_header` and `separator`.

There's more...

We can read and write CSV files in lazy mode with LazyFrame. This lets us utilize the full potential of Polars. You can use `.scan_csv()` instead of `.read_csv()`. Similar parameters to the ones you saw earlier are available in `.scan_csv()` as well:

```
lf = pl.scan_csv('../data/customer_shopping_data_no_header.csv',
             has_header=False,
             new_columns=column_names,
             try_parse_dates=True,
             schema_overrides={'age': pl.Int8, 'quantity': pl.Int32})
lf.head().collect()
```

The preceding code will return the following output:

shape: (5, 10)

invoice_no	customer_id	gender	age	category	quantity	price	payment_method	invoice_date	shopping_mall
str	str	str	i8	str	i32	f64	str	date	str
"I138884"	"C241288"	"Female"	28	"Clothing"	5	1500.4	"Credit Card"	2022-08-05	"Kanyon"
"I317333"	"C111565"	"Male"	21	"Shoes"	3	1800.51	"Debit Card"	2021-12-12	"Forum Istanbul...
"I127801"	"C266599"	"Male"	20	"Clothing"	1	300.08	"Cash"	2021-11-09	"Metrocity"
"I173702"	"C988172"	"Female"	66	"Shoes"	5	3000.85	"Credit Card"	2021-05-16	"Metropol AVM"
"I337046"	"C189076"	"Female"	53	"Books"	4	60.6	"Cash"	2021-10-24	"Kanyon"

Figure 2.7 – The result of using the fetch method

Also, you can write to a CSV file while in lazy mode with `.sink_csv()`. This lets you write the output to a CSV file with streaming mode, allowing you to write a file that is larger than the RAM on your machine:

```
lf.sink_csv('../data/output/shopping_data_output_sink.csv')
```

See also

There are many other useful parameters we couldn't cover in this recipe. Please see other documentation pages for additional information on reading and writing CSV files:

- https://pola-rs.github.io/polars/py-polars/html/reference/api/polars.read_csv.html

- https://pola-rs.github.io/polars/py-polars/html/reference/api/polars.DataFrame.write_csv.html

- https://pola-rs.github.io/polars/py-polars/html/reference/api/polars.scan_csv.html
- https://pola-rs.github.io/polars/py-polars/html/reference/api/polars.LazyFrame.sink_csv.html

Reading and writing Parquet files

The Parquet file format is an open source columnar file format that's efficient for data storage and processing. This column-oriented format is suitable for analytics workloads and efficient compression. The Parquet file format is very common in big data analytics.

In this recipe, you will learn how to read and write Parquet files in both a DataFrame and LazyFrame.

Getting ready

Toward the end of the recipe, you'll need the `pyarrow` library. If you haven't yet installed it, run the following command:

```
>>> pip install pyarrow
```

How to do it...

We'll first cover reading a Parquet file:

1. Read a Parquet file:

```
parquet_input_file_path = '../data/venture_funding_deals.
parquet'
df = pl.read_parquet(
    parquet_input_file_path,
    columns=['Company', 'Amount', 'Valuation', 'Industry'],
    row_index_name='row_cnt'
)
df.head()
```

The preceding code will return the following output:

shape: (5, 4)

Company	Amount	Valuation	Industry
str	str	str	str
"OpenAI"	"$10,000,000,00...	"n/a"	"Artificial int...
"Stripe"	"$6,500,000,000...	"$50,000,000,00...	"Fintech"
"Inflection AI"	"$1,300,000,000...	"$4,000,000,000...	"Artificial int...
"Anthropic"	"$1,250,000,000...	"$4,000,000,000...	"Artificial int...
"Generate Capit...	"$1,030,900,000...	"n/a"	"Energy"

Figure 2.8 – The first five rows of the parquet file

2. Read only the schema of a Parquet file:

```
pl.read_parquet_schema(parquet_input_file_path)
```

The preceding code will return the following output:

```
>> {'Company': String, 'Amount': String, 'Lead investors':
String, 'Valuation': String, 'Industry': String, 'Date
reported': String}
```

3. Write a DataFrame to a Parquet file:

```
parquet_output_file_path = '../data/output/venture_funding_
deals_output.parquet'
df.write_parquet(parquet_output_file_path, compression='lz4',
compression_level=10)
```

4. Read a Parquet file with LazyFrame:

```
lf = pl.scan_parquet(parquet_input_file_path)
lf.head().collect()
```

The preceding code will return the following output:

shape: (5, 6)

Company	Amount	Lead investors	Valuation	Industry	Date reported
str	str	str	str	str	str
"OpenAI"	"$10,000,000,00...	"Microsoft"	"n/a"	"Artificial int...	"1/23/23"
"Stripe"	"$6,500,000,000...	"n/a"	"$50,000,000,00...	"Fintech"	"3/15/23"
"Inflection AI"	"$1,300,000,000...	"Microsoft, Rei...	"$4,000,000,000...	"Artificial int...	"6/29/23"
"Anthropic"	"$1,250,000,000...	"Amazon"	"$4,000,000,000...	"Artificial int...	"9/25/23"
"Generate Capit...	"$1,030,900,000...	"n/a"	"n/a"	"Energy"	"1/6/23"

Figure 2.9 – Scanning the parquet file

5. Write or sink a Parquet file with LazyFrame:

```
lf.sink_parquet(parquet_output_file_path, maintain_order=False)
```

How it works...

The `columns` parameter in `.read_parquet()` lets you choose which columns to read. The `row_index_name` parameter adds a row number column to the DataFrame. There are several other utility parameters to adjust the way you read your Parquet file.

I used `lz4` for the `compression` parameter when using `.write_parquet()`. Using `lz4` compression in particular helps with compression performance. There are a few other options you can choose from depending on your needs such as `snappy` and `zstd`.

Setting the `maintain_order` parameter to `false` helps with a slight performance boost for the `.sink_parquet()` method. This parameter is not available in the `.write_parquet()` method. You can also choose to turn the optimization parameters such as `predicate pushdown` and `projection pushdown` on and off, as well as adjust the compression details as you saw with `.write_parquet()`.

There are many parameters available with these methods. To learn more about additional parameters, I'd recommend reading through the Polars documentation pages that you can find under this recipe's *See also* section.

One important parameter is the `storage_options` parameter in `.read_parquet()` and `.scan_parquet()`. That parameter helps connect to cloud data sources, as we mentioned in the previous recipe, *Reading and writing CSV files*. We'll cover this in *Chapter 11, Working with Common Cloud Data Sources*.

There's more...

If you're used to working with Parquet files, especially large files, you might wonder how you can work with partitioned Parquet files in Polars. The `.read_parquet()` and `.scan_parquet()` methods allow you to read hive partitioned files by setting the `hive_partitioning` parameter to True. You can also read and write partitioned parquet files in Polars with the help of Apache Arrow, which we'll cover in the code below.

The `pyarrow_options` parameter is available both in `.read_parquet()` and `.write_parquet()`. The `pyarrow_options` parameter will use the Parquet reader in the `pyarrow` library instead of the one in Rust. It needs to be used in conjunction with the `use_pyarrow` parameter:

- Read a partitioned Parquet file:

```
partitioned_parquet_input_file_path = '../data/venture_funding_
deals_partitioned'
df = pl.read_parquet(
    partitioned_parquet_input_file_path,
    use_pyarrow=True,
```

```
        pyarrow_options={'partitioning': 'hive'}
    )
    df.head()
```

The preceding code will return the following output:

shape: (5, 6)

Company	Amount	Lead investors	Valuation	Date reported	Industry
str	str	str	str	str	cat
"Restaurant365"	"$135,000,000"	"KKR, L Cattert...	"$1,000,000,000...	"5/19/23"	"Accounting"
"Madhive"	"$300,000,000"	"Goldman Sachs ...	"$1,000,000,000...	"6/13/23"	"Advertising"
"Ursa Major,"	"$100,000,000"	"BlackRock, Spa...	"n/a"	"4/26/23"	"Aerospace"
"Indigo"	"$250,000,000"	"Flagship Pione...	"na"	"9/15/23"	"Agriculture"
"Chronosphere"	"$115,000,000"	"GV"	"n/a"	"1/9/23"	"Analytics"

Figure 2.10 – Read the partitioned parquet file

The default value for partitioning is hive. It's expected that each directory name follows a key-value pair structure such as year=2023. You can see that our partitioned file is in that format:

```
∨ venture_funding_deals_partitioned
  > Industry=Accounting
  > Industry=Advertising
  > Industry=Aerospace
  > Industry=Agriculture
```

Figure 2.11 – How the file is partitioned

For other partitioning flavors other than hive, please refer to this pyarrow documentation page: https://arrow.apache.org/docs/python/generated/pyarrow.parquet.read_table.html.

> **Tip**
>
> In the preceding example, you can omit specifying the pyarrow_options={'partitioning': 'hive'} parameter since it's the default option. Using only use_pyarrow=True allows you to read a partitioned Parquet file.

- Write to a partitioned Parquet file:

```
partitioned_parquet_output_file_path = '../data/output/venture_funding_deals_partitioned_output'
df.write_parquet(
    partitioned_parquet_output_file_path,
```

```
        use_pyarrow=True,
        pyarrow_options={
            'partition_cols': ['Industry'],
            'existing_data_behavior': 'overwrite_or_ignore'
        }
    )
```

You specify partitioning with the partition_cols parameter along with others you can choose from pyarrow's Parquet writer. In the preceding example, I used the overwrite_or_ignore parameter to overwrite files with the same name, ignoring any existing data.

You can specify other parameters inside the pyarrow_options parameter. Refer to these pyarrow documentation pages for additional information:

- https://arrow.apache.org/docs/python/generated/pyarrow.parquet. read_table.html

- https://arrow.apache.org/docs/python/generated/pyarrow.parquet. write_table.html

- https://arrow.apache.org/docs/python/generated/pyarrow.parquet. write_to_dataset.html

The cool thing about Polars is that you can always use pyarrow to help with whatever Polars has not yet implemented natively. We'll cover how Polars works well with other libraries , including pyarrow, in more detail in *Chapter 10, Interoperability with Other Python Libraries.*

See also

To learn more about reading and writing parquet files, please refer to these additional resources:

- https://pola-rs.github.io/polars/py-polars/html/reference/api/ polars.read_parquet.html

- https://pola-rs.github.io/polars/py-polars/html/reference/api/ polars.scan_parquet.html

- https://pola-rs.github.io/polars/py-polars/html/reference/api/ polars.read_parquet_schema.html

- https://pola-rs.github.io/polars/py-polars/html/reference/api/ polars.DataFrame.write_parquet.html

- https://pola-rs.github.io/polars/py-polars/html/reference/api/ . polars.LazyFrame.sink_parquet.html

- https://docs.pola.rs/user-guide/getting-started/#reading-writing

Reading and writing Delta Lake tables

Delta Lake is an open source storage layer built on top of the Parquet format. Delta Lake has more features than the Parquet format such as versioning and ACID guarantees. It's basically a Parquet file with some additional benefits.

Many data pipelines nowadays are built in lakehouse architecture, which is a mix of data lakes and warehouses. Delta Lake table is a popular option and is used by many companies. Delta Lake tables can essentially be stored in your data lake but also be queried and used like relational tables. So, Polars being able to work with Delta Lake tables is a big plus.

In this recipe, we'll look at how to read and write Delta Lake tables with a few useful parameters.

Getting ready

This recipe requires you to install another Python library, `deltalake`. It's a dependency required for Polars to work with Delta Lake tables. Run the following command to install it in your Python environment:

```
>>> pip install deltalake
```

How to do it...

Here are the steps for reading and writing Delta Lake tables:

1. Read a Delta Lake table to a DataFrame:

    ```
    delta_input_file_path = '../data/venture_funding_deals_delta'
    df = pl.read_delta(delta_input_file_path)
    df.head()
    ```

2. Scan a Delta Lake table to a LazyFrame:

    ```
    lf = pl.scan_delta(delta_input_file_path)
    lf.head().collect()
    ```

The preceding steps result in the following:

shape: (5, 6)

Company	Amount	Lead investors	Valuation	Industry	Date reported
str	str	str	str	str	str
"OpenAI"	"$10,000,000,00...	"Microsoft"	"n/a"	"Artificial int...	"1/23/23"
"Stripe"	"$6,500,000,000...	"n/a"	"$50,000,000,00...	"Fintech"	"3/15/23"
"Inflection AI"	"$1,300,000,000...	"Microsoft, Rei...	"$4,000,000,000...	"Artificial int...	"6/29/23"
"Anthropic"	"$1,250,000,000...	"Amazon"	"$4,000,000,000...	"Artificial int...	"9/25/23"
"Generate Capit...	"$1,030,900,000...	"n/a"	"n/a"	"Energy"	"1/6/23"

Figure 2.12 – The DataFrame from a Delta Lake table

3. Write a DataFrame to a Delta Lake table:

```
df.write_delta('../data/output/venture_funding_deals_delta_
output', mode='overwrite')
```

4. Write to a partitioned Delta Lake table:

```
delta_partitioned_output_file_path = '../data/output/venture_
funding_deals_delta_partitioned_output'
delta_write_options = {'partition_by': 'Industry'}
df.write_delta(
    delta_partitioned_output_file_path,
    mode='overwrite',
    delta_write_options=delta_write_options
)
```

5. Read a partitioned Delta Lake table:

```
df = pl.read_delta(delta_partitioned_output_file_path)
df.head()
```

The preceding code will return the following output:

shape: (5, 6)

Company	Amount	Lead investors	Valuation	Industry	Date reported
str	str	str	str	str	str
"SandboxAQ"	"$500,000,000"	"n/a"	"n/a"	"Artifical inte...	"2/14/23"
"Humane"	"$100,000,000"	"Kindred Ventur...	"n/a"	"Artifical inte...	"3/8/23"
"Generate Capit...	"$1,030,900,000...	"n/a"	"n/a"	"Energy"	"1/6/23"
"Our Next Energ...	"$300,000,000"	"Franklin Templ...	"$1,200,000,000...	"Energy"	"2/1/23"
"Ohmium Interna...	"$250,000,000"	"TPG Rise Clima...	"n/a"	"Energy"	"4/26/23"

Figure 2.13 – The DataFrame from a partitioned Delta Lake table

6. Read only a partition:

```
df = pl.read_delta(
    delta_partitioned_output_file_path,
    pyarrow_options={'partitions': [('Industry', '=',
'Accounting')]}
)
df.head()
```

The preceding code will return the following output:

shape: (1, 6)

Company	Amount	Lead investors	Valuation	Industry	Date reported
str	str	str	str	str	str
"Restaurant365"	"$135,000,000"	"KKR, L Cattert...	"$1,000,000,000...	"Accounting"	"5/19/23"

Figure 2.14 – The DataFrame from a partition of a Delta Lake table

How it works...

Although the way you read, scan, and write data in this context is the same as for other file formats, there are a few parameters worth mentioning.

The delta_write_option parameter helps you specify additional configurations such as how you want to partition your data when writing to a file. The pyarrow_options parameter also helps you configure additional settings. It's only available for the .read_delta() and .scan_delta() functions. In one of the preceding examples, I'm specifying a particular partition to read with pyarrow_options.

The mode parameter is handy and determines how you write your new Delta Lake table over your existing Delta Lake table. The merge option would be the one you want to use when you want to update data with a merge or upsert operation. Polars's documentation page gives you a good example of this: https://docs.pola.rs/py-polars/html/reference/api/polars.DataFrame.write_delta.html.

Lastly, the version parameter helps you retrieve a specific version of your Delta Lake table. You'll get the latest version if you don't specify anything.

To learn more about the parameters available for reading and writing Delta Lake tables, please refer to the documentation found under this recipe's *See also* section.

There's more...

Your Delta Lake tables are most likely stored in a cloud storage solution such as an S3 bucket or Azure Data Lake Storage. You can work with cloud storage solutions with the `storage_options` parameter. You'd need your AWS account and an S3 bucket set up to do this. If you're new to S3, you can refer to this documentation from AWS: `https://aws.amazon.com/s3/getting-started/`.

We'll dive into more detail on working with cloud sources in *Chapter 11*, *Working with Common Cloud Data Sources*, but here's a code snippet that reads data from a S3 bucket:

```
table_path = 's3://YOUR_S3BUCKET_URI/YOUR_DELTA_TABLE'
storage_options= {
    'aws_access_key_id': 'YOUR_ACCESS_KEY',
    'aws_secret_access_key': 'YOUR_SECRET_ACCESS_KEY',
    'aws_region': 'YOUR_REGION'
}
df = pl.read_delta(table_path, storage_options=storage_options)
df.head()
```

The preceding code will return the following output:

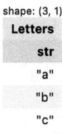

Figure 2.15 – The DataFrame based on the source in S3

> **Note**
> Polars checks for values such as aws access key id and aws secret access key as environment variables. However, you can achieve the same result by providing a dictionary manually through the `storage_options` parameter. I just showed how to do this in the preceding example.

> **Important note**
> Polars credentials, like AWS keys, should never be hardcoded in production. Consider storing them in environmental variables, a secret manager, or a key vault.

See also

To learn more about working with Delta Lake tables, please use these resources:

- `https://pola-rs.github.io/polars/py-polars/html/reference/api/polars.read_delta.html`

- `https://pola-rs.github.io/polars/py-polars/html/reference/api/polars.scan_delta.html`

- `https://pola-rs.github.io/polars/py-polars/html/reference/api/polars.DataFrame.write_delta.html`

- `https://github.com/delta-io/delta-rs/blob/395d48b47ea638b70415899dc035cc895b220e55/python/deltalake/writer.py#L65`

- `https://delta-io.github.io/delta-rs/python/usage.html?highlight=backend#loading-a-delta-table`

Reading and writing JSON files

JavaScript Object Notation (**JSON**) is an open source file format used to store and transport data. It can easily be parsed into a JavaScript object. JSON is language independent and is used in projects with other programming languages that require a lightweight data exchange format. JSON stores and represents data as key-value pairs. In Python terms, JSON is very much like data that is stored in Python dictionaries.

In this recipe, we'll cover how to read and write JSON files in Polars. We'll also cover how to work with a different variation of JSON: **Newline Delimited JSON** (**NDJSON**). It is also called **JSON Lines** (**JSONL**) or **Line-Delimited JSON** (**LDJSON**). As the name suggests, each line is a JSON object.

How to do it...

Next, we'll dive into how to work with JSON files in Polars:

1. Read a JSON file, showing the first 10 columns:

    ```
    df = pl.read_json('../data/world_population.json')
    df.select(df.columns[:10]).head()
    ```

    The preceding code will return the following output:

shape: (5, 10)

place	pop1980	pop2000	pop2010	pop2022	pop2023	pop2030	pop2050	country	area
i64	f64	f64	f64	f64	i64	f64	f64	str	i64
356	6.96828385e8	1.0596e9	1.2406e9	1.4172e9	1428627663	1.5150e9	1.6705e9	"India"	3287590
156	9.82372466e8	1.2641e9	1.3482e9	1.4259e9	1425671352	1.4156e9	1.3126e9	"China"	9706961
840	2.23140018e8	2.82398554e8	3.11182845e8	3.38289857e8	339996563	3.52162301e8	3.75391963e8	"United States"	9372610
360	1.48177096e8	2.14072421e8	2.44016173e8	2.75501339e8	277534122	2.921501e8	3.17225213e8	"Indonesia"	1904569
586	8.0624057e7	1.54369924e8	1.94454498e8	2.35824862e8	240485658	2.74029836e8	3.67808468e8	"Pakistan"	881912

Figure 2.16 – The first five rows of the DataFrame with the first 10 columns

2. Write a DataFrame to a JSON file:

```
df.write_json('../data/output/world_population_output.json')
```

3. Read an NDJSON file. The `.jsonl` file extension comes from JSONL:

```
df = pl.read_ndjson('../data/world_population.jsonl')
df.select(df.columns[:10]).head()
```

The preceding code will return the following output:

shape: (5, 10)

place	pop1980	pop2000	pop2010	pop2022	pop2023	pop2030	pop2050	country	area
i64	f64	f64	f64	f64	i64	f64	f64	str	i64
356	6.96828385e8	1.0596e9	1.2406e9	1.4172e9	1428627663	1.5150e9	1.6705e9	"India"	3287590
156	9.82372466e8	1.2641e9	1.3482e9	1.4259e9	1425671352	1.4156e9	1.3126e9	"China"	9706961
840	2.23140018e8	2.82398554e8	3.11182845e8	3.38289857e8	339996563	3.52162301e8	3.75391963e8	"United States"	9372610
360	1.48177096e8	2.14072421e8	2.44016173e8	2.75501339e8	277534122	2.921501e8	3.17225213e8	"Indonesia"	1904569
586	8.0624057e7	1.54369924e8	1.94454498e8	2.35824862e8	240485658	2.74029836e8	3.67808468e8	"Pakistan"	881912

Figure 2.17 – The first five rows of the DataFrame from the NDJSON file

For those who are not familiar with NDJSON format, the input file looks like this:

```
{"key1": "a", "key2: "b", "key3": 3}
{"key1": "d", "key2: "d", "key3": 6}
{"key1": "f", "key2: "g", "key3": 9}
```

4. Write a DataFrame to an NDJSON file:

```
df.write_ndjson('../data/output/world_population_output.jsonl')
```

How it works...

Both reading and writing JSON and NDJSON files follow the same format. They have the same parameters, including `schema` and `schema_overrides`. With `schema`, you provide a list of both column names and data types. The data provided needs to match the original structure of the dataset. With `schema_overrides`, you can provide the schema for the subset of the columns.

> **Note**
>
> Since Polars version 1.0.0, `pl.read_json()` no longer handles JSON files created by `.serialize()`. Instead, users should use `pl.DataFrame.deserialize()`.
>
> The `.write_json()` method now exclusively writes row-oriented JSON. The parameters row_oriented and pretty have been eliminated. To serialize a DataFrame, users should use `.serialize()`.

There's more...

You can read NDJSON files in LazyFrame utilizing the optimizations of the Polars lazy API:

```
lf = pl.scan_ndjson('../data/world_population.jsonl')
lf.select(lf.columns[:10]).head().collect()
```

The preceding code will return the following output:

shape: (5, 10)

place	pop1980	pop2000	pop2010	pop2022	pop2023	pop2030	pop2050	country	area
i64	f64	f64	f64	f64	i64	f64	f64	str	i64
356	6.96828385e8	1.0596e9	1.2406e9	1.4172e9	1428627663	1.5150e9	1.6705e9	"India"	3287590
156	9.82372466e8	1.2641e9	1.3482e9	1.4259e9	1425671352	1.4156e9	1.3126e9	"China"	9706961
840	2.23140018e8	2.82398554e8	3.11182845e8	3.38289857e8	339996563	3.52162301e8	3.75391963e8	"United States"	9372610
360	1.48177096e8	2.14072421e8	2.44016173e8	2.75501339e8	277534122	2.921501e8	3.17225213e8	"Indonesia"	1904569
586	8.0624057e7	1.54369924e8	1.94454498e8	2.35824862e8	240485658	2.74029836e8	3.67808468e8	"Pakistan"	881912

Figure 2.18 – Read the NDJSON file into a LazyFrame

You can learn more about `.scan_jdjson()` from the documentation page, which can be found here: https://pola-rs.github.io/polars/py-polars/html/reference/api/polars.scan_ndjson.html.

See also

To learn more about working with JSON files in Polars, please refer to these additional resources:

- https://pola-rs.github.io/polars/py-polars/html/reference/api/polars.read_json.html

- https://pola-rs.github.io/polars/py-polars/html/reference/api/polars.DataFrame.write_json.html

- https://pola-rs.github.io/polars/py-polars/html/reference/api/polars.read_ndjson.html

- https://pola-rs.github.io/polars/py-polars/html/reference/api/
  polars.DataFrame.write_ndjson.html
- https://pola-rs.github.io/polars/py-polars/html/reference/api/
  polars.scan_ndjson.html

Reading and writing Excel files

We all know that Excel is one of the most popular data analysis tools out there. It still is the one that most of us are familiar with. Being able to work with Excel in Polars is essential for data analysts. In this recipe, we'll go through reading and writing Excel files, as well as utilizing some of their useful parameters.

Getting ready

This recipe requires a few Python libraries on top of Polars. You can install it with the following command:

```
>>> pip install xlsx2csv xlsxwriter
```

How to do it...

We'll cover how to read and write Excel files using the following steps:

1. Let's first read a CSV file into a DataFrame and write it to an Excel file:

    ```
    output_file_path = '../data/output/financial_sample_output.xlsx'

    df = pl.read_csv('../data/customer_shopping_data.csv')
    df.write_excel(
        output_file_path,
        worksheet='Output Sheet1',
        header_format={'bold': True}
    )
    ```

2. Read an Excel file to a DataFrame:

    ```
    input_file_path = '../data/output/customer_shopping_data.xlsx'
    df = pl.read_excel(
        input_file_path,
        sheet_name='Output Sheet1',
        engine='xlsx2csv',
        read_options={'try_parse_dates': True}
    )
    df.head()
    ```

The preceding code will return the following output:

shape: (5, 10)

invoice_no	customer_id	gender	age	category	quantity	price	payment_method	invoice_date	shopping_mall
str	str	str	i64	str	i64	f64	str	date	str
"I138884"	"C241288"	"Female"	28	"Clothing"	5	1500.4	"Credit Card"	2022-08-05	"Kanyon"
"I317333"	"C111565"	"Male"	21	"Shoes"	3	1800.51	"Debit Card"	2021-12-12	"Forum Istanbul…
"I127801"	"C266599"	"Male"	20	"Clothing"	1	300.08	"Cash"	2021-11-09	"Metrocity"
"I173702"	"C988172"	"Female"	66	"Shoes"	5	3000.85	"Credit Card"	2021-05-16	"Metropol AVM"
"I337046"	"C189076"	"Female"	53	"Books"	4	60.6	"Cash"	2021-10-24	"Kanyon"

Figure 2.19 – The first five rows of the DataFrame

How it works...

First of all, you needed the `xlsx2csv` library because we needed to specify to parse a date column upon reading the file. You can change the engine to either `openpyxl` or `calamine` if you'd like, by utilizing the `engine` parameter. By default, Polars uses `calamine` as it's the fastest option than other engines.

A few of the parameters that come in handy are `sheet_id` and `sheet_name`. They allow you to only read particular sheets in your Excel file. If you're reading multiple sheets, then `.read_excel()` returns a Python dictionary containing the sheet names and associated DataFrames.

Another parameter I used that's worth mentioning is the `read_options` parameter. It is available because together with `xlsx2csv`, it converts the file to CSV first. Then use `.csv_read()` to parse the data. This way, you get to use the parameters that are available in the CSV reader function as well.

When writing out a DataFrame to an Excel file, you have plenty of parameters you can tweak to adjust the output. You can even add column or row totals and change the style of the header.

There are many other parameters you can utilize according to your needs for reading and writing Excel files. I won't cover everything, but there are Polars documentation pages to learn more, which are listed under this recipe's *See also* section.

See also

If you'd like to learn more about working with Excel files, please consult these resources:

- `https://pola-rs.github.io/polars/py-polars/html/reference/api/polars.read_excel.html`

- `https://pola-rs.github.io/polars/py-polars/html/reference/api/polars.DataFrame.write_excel.html`

Reading and writing other data file formats

There are many other formats aside from the ones introduced earlier. Polars keeps adding features to work with more formats in its frequent updates. We'll be going over a few other data file formats to read from and write to in Polars.

In this recipe, we'll cover reading from and/or writing to the Arrow IPC format, Apache Avro, and Apache Iceberg.

These file formats are not as common as the other ones we covered in earlier recipes. However, there are still use cases where companies and people need to work with these formats.

Getting ready

You'll need to install a few other libraries other than Polars for this recipe. They are `pyiceberg`, `numpy`, and `pyarrow`. Run the following commands in Terminal to install them if you haven't already:

```
>>> pip install pyiceberg
>>> pip install numpy
>>> pip install pyarrow
```

How to do it...

Here are the steps for working with other data formats in Polars:

1. Let's create a DataFrame from the CSV file we used earlier and write to an Arrow IPC format (Feather V2):

   ```
   csv_input_file_path = '../data/customer_shopping_data.csv'
   ipc_file_path = '../data/output/customer_shopping_data.arrow'
   df = pl.read_csv(csv_input_file_path)
   df.write_ipc(ipc_file_path)
   ```

2. Read Arrow IPC to a DataFrame:

   ```
   df = pl.read_ipc(ipc_file_path)
   df.head()
   ```

3. Create a file in Avro format from a JSON file, selecting only three columns:

   ```
   avro_file_path = '../data/world_population.avro'
   df = (
       pl.read_json('../data/world_population.json')
       .select(['country', 'pop2023', 'density'])
   )
   df.write_avro(avro_file_path)
   ```

4. Read Avro to a DataFrame:

```
df = pl.read_avro(avro_file_path)
df.head()
```

All of the output of the preceding DataFrames should look like the following:

shape: (5, 3)

country	pop2023	density
str	i64	f64
"India"	1428627663	480.5033
"China"	1425671352	151.2696
"United States"	339996563	37.1686
"Indonesia"	277534122	147.8196
"Pakistan"	240485658	311.9625

Figure 2.20 – The first five rows of the Avro file

5. Scan an Iceberg table through its metadata file. We're using a different dataset that I created beforehand:

```
lf = pl.scan_iceberg('../data/my_iceberg_catalog/demo.db/
my_table/metadata/00001-7ad1e6e8-7a0d-4455-ac6d-bcca5a45b494.
metadata.json')
lf.head().collect()
```

The preceding code will return the following output:

shape: (3, 3)

a	b	c
i64	i64	i64
1	4	7
2	5	8
3	6	9

Figure 2.21 – Reading an Iceberg table through its metadata file

How it works...

The syntax is the same when we work with any file format. Polars uses keywords such as `read`, `write`, `scan`, and `sink`.

As for working with Iceberg tables, there are two ways to read them using the `.scan_iceberg()` method. One way is by accessing data through an Iceberg catalog. The other way is by reading its metadata file, which we demonstrated in *step 5*. The former is the recommended way. If you're interested in that approach, you can find a Python script that shows how to create a SQL catalog backed by Postgres database, add a table, and read it in Polars at `https://github.com/PacktPublishing/Polars-Cookbook/blob/main/Chapter02/create_iceberg_catalog_and_table.py`. It is exactly what I used to produce the data for *step 5*.

At the time of writing, only the `scan` method can be used for working with Apache Iceberg.

You can refer to their documentation page to learn more about Apache Iceberg at `https://py.iceberg.apache.org/`.

There's more...

You can read from and write to Arrow IPC with LazyFrame with `.scan_ipc()` and `.sink_ipc()`:

- Scan Arrow IPC:

```
lf = pl.scan_ipc(ipc_file_path)
```

- Sink Arrow IPC:

```
lf.sink_ipc('../data/output/customer_shopping_data.arrow')
```

  The preceding code will return the following output:

```
>> InvalidOperationError: sink_Ipc(IpcWriterOptions {
compression: Some(ZSTD), maintain_order: true }) not yet
supported in standard engine. Use 'collect().write_parquet()'
```

  Since `.scan_ipc()` is not yet supported in the streaming engine at the time of writing, it throws an error.

  You can still output an Arrow IPC using `.write_ipc()` or read in a different way, such as by using `.scan_csv()`, for example.

- Scan Arrow and then write with `.write_ipc()`:

```
lf.collect().write_ipc('../data/output/customer_shopping_data_
lazy.arrow')
```

- Scan a CSV file and then sink it to Arrow:

```
lf = pl.scan_csv(csv_input_file_path)
lf.sink_ipc('../data/output/customer_shopping_data_from_csv.
arrow')
```

Also, you can work with the Arrow IPC record batch stream using `.read_ipc_stream()` and `.write_ipc_stream()`. You can find more information on these documentation pages at the following addresses:

- `https://pola-rs.github.io/polars/py-polars/html/reference/api/polars.read_ipc_stream.html`

- `https://pola-rs.github.io/polars/py-polars/html/reference/api/polars.DataFrame.write_ipc_stream.html`

- `https://jorisvandenbossche.github.io/arrow-docs-preview/html-option-1/python/ipc.html`

See also

To learn more about working with various data formats, please refer to these resources:

- `https://pola-rs.github.io/polars/py-polars/html/reference/api/polars.read_ipc.html`

- `https://pola-rs.github.io/polars/py-polars/html/reference/api/polars.DataFrame.write_ipc.html`

- `https://pola-rs.github.io/polars/py-polars/html/reference/api/polars.scan_ipc.html`

- `https://pola-rs.github.io/polars/py-polars/html/reference/api/polars.LazyFrame.sink_ipc.html`

- `https://pola-rs.github.io/polars/py-polars/html/reference/api/polars.read_avro.html`

- `https://pola-rs.github.io/polars/py-polars/html/reference/api/polars.DataFrame.write_avro.html`

Reading and writing multiple files

When working on actual data projects, there are cases where data is split into multiple files in a directory. Dealing with each file one by one can be a pain and may distract you from working on other critical components of your project.

In this recipe, we'll cover reading multiple files into a single DataFrame or into multiple DataFrames, as well as writing a DataFrame to multiple files.

How to do it...

Here are some ways to work with multiple files:

1. Write a DataFrame to multiple CSV files:

    A. Create a DataFrame:

    ```
    data = {'Letter': ['A','B','C'], 'Value': [1,2,3]}
    df = pl.DataFrame(data)
    ```

    B. Split it into multiple DataFrames:

    ```
    dfs = df.group_by(['Letter'])
    print(dfs)
    ```

    The preceding code will return the following output:

    ```
    >> <polars.dataframe.group_by.GroupBy object at 0x154373390>
    ```

    C. Write them to CSV files:

    ```
    for name, df in dfs:
        df.write_csv(f'../data/output/letter_{name[0]}.csv')
    ```

2. Read multiple CSV files into a list of DataFrames:

    ```
    df = pl.read_csv('../data/output/letter_*.csv')
    df.head()
    ```

3. Scan multiple files to LazyFrames:

    ```
    lf = pl.scan_csv('../data/output/letter_*.csv')
    lf.head().collect()
    ```

The output of *steps 2* and *3* will be the following:

shape: (3, 2)

Letter	Value
str	i64
"A"	1
"B"	2
"C"	3

Figure 2.22 – A DataFrame from multiple files

How it works...

`.group_by()` is typically used with aggregations, which we'll cover in more detail in *Chapter 4 Data transformation Techniques*. However, since the object created by `.group_by()` returns is iterable, you can loop through it and get a DataFrame and its name for each group.

When reading multiple files, "*" can be used to specify multiple files. Doing this helps you select files dynamically. In the preceding example, I selected all the files that start with `letter_` and end with `.csv`. This also works with scan functions such as `.scan_csv()`.

There's more...

You can also read multiple files into separate DataFrames while still utilizing Polars's parallelization. `.collect_all()` does the magic of processing multiple files in parallel:

```
import glob
lfs = [pl.scan_csv(file) for file in glob.glob('../data/output/
letter_*.csv')]
dfs = pl.collect_all(lfs)
dfs
```

The preceding code will return the following output:

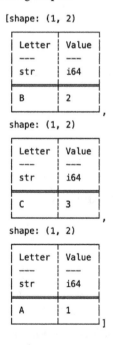

Figure 2.23 – A list of DataFrames

See also

To learn more about working with multiple files, please refer to these additional resources:

- `https://pola-rs.github.io/polars/py-polars/html/reference/dataframe/api/polars.DataFrame.group_by.html#polars.DataFrame.group_by`

- `https://pola-rs.github.io/polars/user-guide/io/multiple/`

Working with databases

In addition to working with files, it's very common to work with databases. Polars has integrations with various databases, whether they're hosted in the cloud or on-premises. Once you understand how to work with one database in Polars, you can apply the same patterns to various other databases.

In this recipe, we'll specifically look at ways to read from and write to a popular database: Postgres. We'll look at how to work with cloud databases in *Chapter 11*, *Working with Common Cloud Data Sources*.

Getting ready

This recipe requires a few additional dependencies. You'll need to install the following libraries:

- `connectorx`

- `adbc-driver-postgresql`

- `pyarrow`

- `pg8000` (or `psycopg2`)

You'll also need to have a Postgres database on your local machine. You can refer to the following websites for more information on how to install a Postgres database locally:

- `https://www.postgresql.org/download/`

- `https://hub.docker.com/_/postgres`

We'll be using a table containing sample data, which I created using the following SQL code in the Postgres database:

```
CREATE TABLE cars (
    brand VARCHAR(255),
    model VARCHAR(255),
    year INT
);
```

```
INSERT INTO cars (brand, model, year)
VALUES
    ('Volvo', 'p1800', 1968),
    ('BMW', 'M1', 1978),
    ('Toyota', 'Celica', 1975);
```

The source for this data is in a tutorial on the W3School website at `https://www.w3schools.com/postgresql/postgresql_create_table.php`.

Also, please note that I stored my Postgres credentials in variables in another Python file, `config.py`.

How to do it...

We'll first look at reading from a database and then writing to a database:

1. Read from Postgres with the `connentorx` default engine:

    ```
    from config import postgres_pass, postgres_user

    uri = f'postgres://{postgres_user}:{postgres_pass}@
    localhost:5432/postgres'
    query = 'SELECT * FROM sandbox.cars'
    df = pl.read_database_uri(query, uri)
    df.head()
    ```

2. Read from Postgres with the **Arrow Database Connectivity (ADBC)** driver:

    ```
    df = pl.read_database_uri(query, uri, engine='adbc')
    df.head()
    ```

3. Read from Postgres using SQLAlchemy:

    ```
    from sqlalchemy import create_engine

    con_string = f'postgresql+pg8000://{postgres_user}:{postgres_
    pass}@localhost:5432/postgres'
    engine = create_engine(con_string)
    conn = engine.connect()

    df = pl.read_database(query, connection=conn)
    df.head()
    ```

Steps 1, *2*, and *3*, will result in the following output:

shape: (3, 3)

brand	model	year
str	str	i32
"Volvo"	"p1800"	1968
"BMW"	"M1"	1978
"Toyota"	"Celica"	1975

Figure 2.24 – The cars table

4. Write a DataFrame to a database:

```
df.write_database(table_name="sandbox.cars_output",
connection=uri, engine="adbc", if_table_exists='replace')
```

5. Write a DataFrame to a database using SQLAlchemy:

```
df.write_database(table_name="sandbox.cars_output",
connection=con_string, engine="sqlalchemy", if_table_
exists='replace')
```

Make sure that the table we created has the expected values:

```
df = pl.read_database_uri('select * from sandbox.cars_output',
uri, engine='adbc')
df.head()
```

The preceding code will return the same output as we saw in *Figure 2.23*.

How it works...

Both .read_database_uri() and .read_database() allow you to read from a database. They have different parameters and options specific to their engine. The rule of thumb is that when you would like to use a connection string or **Unique Resource Identifier** (**URI**), you need to use .read_database_uri(). If you want to use a connection engine created by a library such as SQLAlchemy, then use .read_database(). To learn more about what engine is available to use, refer to the resources under the *See also* section of this recipe.

In the case of using SQLAlchemy, there are two ways to connect to Postgres. The default option is using the psycopg2 Python library. The other one is using the pg8000 Python library. You would need either library to connect.

See also

Please refer to these resources for additional information about reading from and writing to databases:

- `https://pola-rs.github.io/polars/user-guide/io/database/`
- `https://pola-rs.github.io/polars/py-polars/html/reference/api/polars.read_database.html`
- `https://pola-rs.github.io/polars/py-polars/html/reference/api/polars.read_database_uri.html`
- `https://pola-rs.github.io/polars/py-polars/html/reference/api/polars.DataFrame.write_database.html`
- `https://docs.sqlalchemy.org/en/20/core/engines.html`

3

An Introduction to Data Analysis in Python Polars

Data analysis is a broad term that encompasses various steps of inspecting, transforming, and understanding data in order to uncover valuable insights. This chapter focuses on teaching you the fundamentals of data analysis in Python Polars while exploring the dataset. You'll learn how to inspect your data, generate its summary statistics, adjust data types to suit your needs, and clean the data for further analysis.

In this chapter, we're going to cover the following main topics:

- Inspecting the DataFrame
- Casting data types
- Handling duplicate values
- Masking sensitive data
- Visualizing data using Plotly
- Detecting and handling outliers

Technical requirements

You can download the datasets and code from the GitHub repository.

- Datasets: `https://github.com/PacktPublishing/Polars-Cookbook/tree/main/data`
- Code: `https://github.com/PacktPublishing/Polars-Cookbook/tree/main/Chapter03`

It is assumed that you have installed the Polars library in your Python environment:

```
>>> pip install polars
```

Also, it is assumed that you've imported it into your code:

```
import polars as pl
```

Inspecting the DataFrame

The first thing you need to do is to understand what your data looks like. Understanding things such as the schema and basic statistics of your data helps you see the current state of your data. Only then will you be able to see what data operations you'd need to make your data clean and organized for analysis.

In this recipe, we'll cover ways to return a few rows from the dataset, as well as learning how to generate summary statistics and check the column quality.

How to do it...

We'll first read the data and start exploring the DataFrame:

1. Read the dataset:

   ```
   df = pl.read_csv('../data/covid_19_deaths.csv')
   ```

2. Display the first five rows:

   ```
   df.head(5)
   ```

   The preceding code will return the following output:

shape: (5, 16)

Data As Of	Start Date	End Date	Group	Year	Month	State	Sex	Age Group	COVID-19 Deaths	Total Deaths	Pneumonia Deaths	Pneumonia and COVID-19 Deaths	Influenza Deaths	Pneumonia, Influenza, or COVID-19 Deaths	Footnote
str	str	str	str	str	str	str	str	str	i64	i64	i64	i64	i64	i64	str
"09/27/2023"	"01/01/2020"	"09/23/2023"	"By Total"	null	null	"United States"	"All Sexes"	"All Ages"	1146774	12303399	1162844	569264	22229	1760095	null
"09/27/2023"	"01/01/2020"	"09/23/2023"	"By Total"	null	null	"United States"	"All Sexes"	"Under 1 year"	519	73213	1056	95	64	1541	null
"09/27/2023"	"01/01/2020"	"09/23/2023"	"By Total"	null	null	"United States"	"All Sexes"	"0-17 years"	1696	130970	2961	424	509	4716	null
"09/27/2023"	"01/01/2020"	"09/23/2023"	"By Total"	null	null	"United States"	"All Sexes"	"1-4 years"	285	14299	692	66	177	1079	null
"09/27/2023"	"01/01/2020"	"09/23/2023"	"By Total"	null	null	"United States"	"All Sexes"	"5-14 years"	509	22008	818	143	219	1390	null

Figure 3.1 – The first five rows of the DataFrame

3. Display the last *n* rows:

   ```
   df.tail(5)
   ```

The preceding code will return the following output:

shape: (5, 16)

Data As Of	Start Date	End Date	Group	Year	Month	State	Sex	Age Group	COVID-19 Deaths	Total Deaths	Pneumonia Deaths	Pneumonia and COVID-19 Deaths	Influenza Deaths	Pneumonia, Influenza, or COVID-19 Deaths	Footnote
str	str	str	str	str	str	str	str	str	i64	i64	i64	i64	i64	i64	str
"09/27/2023"	"09/01/2023"	"09/23/2023"	"By Month"	"2023"	"9"	"Puerto Rico"	"Female"	"50-64 years"	null	75	14	null	0	14	"One or more da..."
"09/27/2023"	"09/01/2023"	"09/23/2023"	"By Month"	"2023"	"9"	"Puerto Rico"	"Female"	"55-64 years"	0	65	10	0	0	10	null
"09/27/2023"	"09/01/2023"	"09/23/2023"	"By Month"	"2023"	"9"	"Puerto Rico"	"Female"	"65-74 years"	null	91	null	null	0	null	"One or more da..."
"09/27/2023"	"09/01/2023"	"09/23/2023"	"By Month"	"2023"	"9"	"Puerto Rico"	"Female"	"75-84 years"	null	211	36	null	0	38	"One or more da..."
"09/27/2023"	"09/01/2023"	"09/23/2023"	"By Month"	"2023"	"9"	"Puerto Rico"	"Female"	"85 years and o..."	null	265	42	null	null	44	"One or more da..."

Figure 3.2 – The last five rows

4. Display the preview, which provides detailed information on the DataFrame. It shows column details per line, including the column name, data type, and the first few values. This could also be useful as an alternative to the `.head()` and `.tail()` methods:

```
df.glimpse(max_items_per_column=3)
```

The preceding code will return the following output:

```
Rows: 137700
Columns: 16
$ Data As Of                                <str> '09/27/2023', '09/27/2023', '09/27/2023'
$ Start Date                                <str> '01/01/2020', '01/01/2020', '01/01/2020'
$ End Date                                  <str> '09/23/2023', '09/23/2023', '09/23/2023'
$ Group                                     <str> 'By Total', 'By Total', 'By Total'
$ Year                                      <str> None, None, None
$ Month                                     <str> None, None, None
$ State                                     <str> 'United States', 'United States', 'United States'
$ Sex                                       <str> 'All Sexes', 'All Sexes', 'All Sexes'
$ Age Group                                 <str> 'All Ages', 'Under 1 year', '0-17 years'
$ COVID-19 Deaths                           <i64> 1146774, 519, 1696
$ Total Deaths                              <i64> 12303399, 73213, 130970
$ Pneumonia Deaths                          <i64> 1162844, 1056, 2961
$ Pneumonia and COVID-19 Deaths             <i64> 569264, 95, 424
$ Influenza Deaths                          <i64> 22229, 64, 509
$ Pneumonia, Influenza, or COVID-19 Deaths  <i64> 1760095, 1541, 4716
$ Footnote                                  <str> None, None, None
```

Figure 3.3 – The result of the glimpse method

5. Check the estimated size of the DataFrame:

```
df.estimated_size('mb')
```

The preceding code will return the following output:

```
>> 26.869342803955078
```

6. Generate summary statistics of the DataFrame:

```
import polars.selectors as cs
df.select(cs.numeric()).describe()
```

The preceding code will return the following output:

shape: (9, 7)

statistic	COVID-19 Deaths	Total Deaths	Pneumonia Deaths	Pneumonia and COVID-19 Deaths	Influenza Deaths	Pneumonia, Influenza, or COVID-19 Deaths
str	f64	f64	f64	f64	f64	f64
"count"	98270.0	118191.0	92836.0	100816.0	111012.0	93467.0
"null_count"	39430.0	19509.0	44864.0	36884.0	26688.0	44233.0
"mean"	313.586547	2841.952585	336.597085	152.513411	5.002468	505.491778
"std"	5992.341375	56201.384331	6126.573599	2980.886938	110.606691	9256.951591
"min"	0.0	0.0	0.0	0.0	0.0	0.0
"25%"	0.0	43.0	0.0	0.0	0.0	0.0
"50%"	0.0	153.0	18.0	0.0	0.0	25.0
"75%"	50.0	657.0	74.0	21.0	0.0	107.0
"max"	1.146774e6	1.2303399e7	1.162844e6	569264.0	22229.0	1.760095e6

Figure 3.4 – The result of the describe method

7. Display the count of null values:

```
df.null_count()
```

The preceding code will return the following output:

shape: (1, 16)

Data As Of	Start Date	End Date	Group	Year	Month	State	Sex	Age Group	COVID-19 Deaths	Total Deaths	Pneumonia Deaths	Pneumonia and COVID-19 Deaths	Influenza Deaths	Pneumonia, Influenza, or COVID-19 Deaths	Footnote
u32	u32	u32	u32	u32	u32	u32	u32	u32	u32	u32	u32	u32	u32	u32	u32
0	0	0	0	2754	13770	0	0	0	39430	19509	44864	36884	26688	44233	39804

Figure 3.5 – The result of the null_count method

How it works...

Polars has a number of methods that help you inspect and better understand your DataFrames. The .head() and .tail() methods display the first and last *n* rows of the DataFrame. They help you take the first glance at the dataset.

The .glimpse() method gives you detailed information about the DataFrame. The layout is designed to present wide DataFrames in a neat and organized manner. Just like .head() and .tail(), .glimpse() includes data types and the shape of your DataFrame, which we could get with .dtypes and .shape. On top of that, .glimpse() also shows the first few rows of data.

The .estimated_size() method shows the estimated size of the DataFrame. You can specify in which format you want to show the size, such as b, kb, mb, and gb.

The .describe() method displays the summary statistics for each column. In the preceding example, I only selected numeric columns. It not only shows common calculations such as min, max, and mean, but also percentiles and standard deviation. This method helps you see the distribution of your data.

The `.null_count()` method generates a DataFrame that displays the count of null values for each column.

> **Note**
>
> LazyFrames don't have these methods available. When you work in a LazyFrame, you don't have all the information about the dataset because LazyFrames don't process data immediately unlike DataFrames.

There's more...

Here's something that might help you make your data analysis workflows better. Python Polars allows you to customize configurations on how you display your data, preset some parameters upfront, and so on.

One thing I find myself using consistently is `pl.Config.set_tbl_cols()`. This doesn't matter much in Jupyter Notebook environments; however, when you work in a generic Python file or script, you print the result, and it only shows a certain number of columns. But with `pl.Config.set_tbl_cols()`, I can display all the necessary columns I'd like to see:

1. With `print()`, it shows the limited number of columns by default:

    ```
    print(df.head())
    ```

    The preceding code will return the following output:

shape: (5, 16)

Data As Of	Start Date	End Date	Group	...	Pneumonia and COVID-19 Deaths	Influenza Deaths	Pneumonia , Influen za, or COVID-1…	Footnote
---	---	---	---		---	---	---	---
str	---	---	str		i64	---	za, or	str
	str	str				i64	COVID-1…	
							---	
					i64		i64	
09/27/2023	01/01/202 0	09/23/202 3	By Total	...	569264	22229	1760095	null
09/27/2023	01/01/202 0	09/23/202 3	By Total	...	95	64	1541	null
09/27/2023	01/01/202 0	09/23/202 3	By Total	...	424	509	4716	null
09/27/2023	01/01/202 0	09/23/202 3	By Total	...	66	177	1079	null
09/27/2023	01/01/202 0	09/23/202 3	By Total	...	143	219	1390	null

Figure 3.6 – The output when printing the DataFrame

2. Adjust the number of columns to show. If you want to modify the configuration just for a subset of commands, you can use a context manager using a `with` statement:

```
with pl.Config() as config:
    config.set_tbl_cols(11)
    print(df.head(2))
```

Or if you'd like to set the configurations for all the subsequent commands, you can use the following:

```
pl.Config.set_tbl_cols(11)
print(df.head(2))
```

The preceding code will return the following output:

```
shape: (2, 16)
```

Data As Of --- str	Start Date --- str	End Date --- str	Group --- str	Year --- str	Month --- str	…	Pneumo nia Deaths --- i64	Pneumo nia and COVID- 19 Deaths --- i64	Influe nza Deaths --- i64	Pneumo nia, Influe nza, or COV ID-1… --- i64	Footn ote --- str
09/27/ 2023	01/01/ 2020	09/23/ 2023	By Total	null	null	…	116284 4	569264	22229	176009 5	null
09/27/ 2023	01/01/ 2020	09/23/ 2023	By Total	null	null	…	1056	95	64	1541	null

Figure 3.7 – Showing more column names

> **Tip**
> In addition, if you have many configurations you want to use across many Polars projects or scripts, you can save your configurations in a JSON file using `pl.Config.save()` or `pl.Config.save_to_file()`. The former saves the current configurations in a variable as a string and the latter saves your configurations to a JSON file. You can load them for use with `pl.Config.load()` and `pl.Config.load_from_file()`.

See also

Please refer to these additional resources:

- `https://pola-rs.github.io/polars/py-polars/html/reference/dataframe/descriptive.html`
- `https://pola-rs.github.io/polars/py-polars/html/reference/config.html`

Casting data types

When reading your dataset, Polars auto-infers data types and it works well. However, there are cases where Polars doesn't get the data types right or the formatting of a column makes it hard for Polars to infer its data type.

One of the techniques that is useful in this type of situation is to cast or convert a data type to another. In this recipe, we'll cover how to do that.

How to do it...

Here are the steps you can use to cast data types:

1. Check which column you want to use to cast the data type to another:

```
df = pl.read_csv('../data/covid_19_deaths.csv')
df.head()
```

The preceding code will return the following output:

shape: (5, 16)

Data As Of	Start Date	End Date	Group	Year	...	Pneumonia Deaths	Pneumonia and COVID-19 Deaths	Influenza Deaths	Pneumonia, Influenza, or COVID-19 Deaths	Footnote
str	str	str	str	str	...	i64	i64	i64	i64	str
"09/27/2023"	"01/01/2020"	"09/23/2023"	"By Total"	null	...	1162844	569264	22229	1760095	null
"09/27/2023"	"01/01/2020"	"09/23/2023"	"By Total"	null	...	1056	95	64	1541	null
"09/27/2023"	"01/01/2020"	"09/23/2023"	"By Total"	null	...	2961	424	509	4716	null
"09/27/2023"	"01/01/2020"	"09/23/2023"	"By Total"	null	...	692	66	177	1079	null
"09/27/2023"	"01/01/2020"	"09/23/2023"	"By Total"	null	...	818	143	219	1390	null

Figure 3.8 – The first five rows of the COVID dataset

2. Both Date and `Year` columns are read as string. Let's work on casting those data types to correct ones.

3. Cast data types of certain columns of the DataFrame:

```
df.with_columns(
    pl.col('Data As Of').str.strptime(pl.Date, '%m/%d/%Y'),
    pl.col('Start Date').str.strptime(pl.Date, '%m/%d/%Y'),
    pl.col('End Date').str.strptime(pl.Date, '%m/%d/%Y'),
    pl.col('End Date').str.to_date('%m/%d/%Y').alias('End Date 2'),
    pl.col('Year').cast(pl.Int64)
).head()
```

The preceding code will return the following output:

shape: (5, 17)

Data As Of	Start Date	End Date	Group	Year	Month	State	Sex	Age Group	COVID-19 Deaths	Total Deaths	Pneumonia Deaths	Pneumonia and COVID-19 Deaths	Influenza Deaths	Pneumonia, Influenza, or COVID-19 Deaths	Footnote	End Date 2
date	date	date	str	i64	str	str	str	str	i64	i64	i64	i64	i64	i64	str	date
2023-09-27	2020-01-01	2023-09-23	"By Total"	null	null	"United States"	"All Sexes"	"All Ages"	1146774	12303399	1162844	569264	22229	1760095	null	2023-09-23
2023-09-27	2020-01-01	2023-09-23	"By Total"	null	null	"United States"	"All Sexes"	"Under 1 year"	519	73213	1056	95	64	1541	null	2023-09-23
2023-09-27	2020-01-01	2023-09-23	"By Total"	null	null	"United States"	"All Sexes"	"0-17 years"	1696	130970	2961	424	509	4716	null	2023-09-23
2023-09-27	2020-01-01	2023-09-23	"By Total"	null	null	"United States"	"All Sexes"	"1-4 years"	285	14299	692	66	177	1079	null	2023-09-23
2023-09-27	2020-01-01	2023-09-23	"By Total"	null	null	"United States"	"All Sexes"	"5-14 years"	509	22008	818	143	219	1390	null	2023-09-23

Figure 3.9 – Data types adjusted

Note that the preceding code doesn't overwrite the changes in place. If you want to keep the updated DataFrame, then you can simply assign a new variable:

```
updated_df = (
df.with_columns(
    pl.col('Data As Of').str.strptime(pl.Date, '%m/%d/%Y'),
    pl.col('Start Date').str.strptime(pl.Date, '%m/%d/%Y'),
    pl.col('End Date').str.strptime(pl.Date, '%m/%d/%Y'),
    pl.col('End Date').str.to_date('%m/%d/%Y').alias('End Date 2'),
    pl.col('Year').cast(pl.Int64)
    )
)
```

4. Cast date types with a LazyFrame:

```
lf = pl.scan_csv('../data/covid_19_deaths.csv')
lf.with_columns(
    pl.col('Data As Of').str.strptime(pl.Date, '%m/%d/%Y'),
    pl.col('Start Date').str.strptime(pl.Date, '%m/%d/%Y'),
    pl.col('End Date').str.strptime(pl.Date, '%m/%d/%Y'),
    pl.col('End Date').str.to_date('%m/%d/%Y').alias('End Date 2'),
    pl.col('Year').cast(pl.Int64)
).collect().head()
```

How it works...

We used the `.with_columns()` method to overwrite the data types. Just as a refresher, in Polars, once we're inside `.with_columns()`, we're in the world of expressions. This is where all the operations available in Polars' expressions API shine.

For date columns, we used `.str.strptime()` to convert from the string data type to a date. You can also use other expressions, such as `.str.to_date()` and `.str.to_datetime()`, to convert columns to a date or datetime. When Polars can't infer the correct date/datetime data types, it uses the `Utf8` data type, which is a UTF-8-encoded string. Hence, the use of the `str` namespace comes into play to convert from a string to date/datetime.

The `.cast()` expression is also commonly used to cast data types. You can do conversions such as integer to float, float to integer, integer to string, and date/datetime to integer. You can refer to the Polars documentation page to learn more: `https://pola-rs.github.io/polars/user-guide/expressions/casting/`.

> **Tip**
>
> The `strict` parameter in the `.cast()` expression can determine how Polars behaves when Polars cannot convert the original data type to the target data type. The default setting of the parameter is `strict=True`, which results in Polars issuing an error to inform the user about unsuccessful conversions. Conversely, with `strict=False`, any values that cannot be converted to the desired data type will be changed to null without any warning.

There's more...

You can use parameters such as `try_parse_dates`, `dtypes`, and `schema` to specify data types when reading data as opposed to casting data types after reading the data in a DataFrame/LazyFrame.

There is another thing that's worth mentioning about data type casting: the conversion of time zones. Time zones are something that can give you a headache in your data analysis workflows. The good news is that Polars has built-in expressions that you can utilize to facilitate the conversion of time zones. You can set a time zone using the `.dt.replace_time_zone` expression on a datetime column. Also, you can change the time zone of your datetime column with the `.dt.convert_time_zone` expression.

See also

For more information on casting data types, please refer to these additional resources:

- `https://pola-rs.github.io/polars/user-guide/expressions/casting/`
- `https://pola-rs.github.io/polars/py-polars/html/reference/expressions/api/polars.Expr.cast.html`
- `https://pola-rs.github.io/polars/py-polars/html/reference/expressions/api/polars.Expr.dt.replace_time_zone.html`
- `https://pola-rs.github.io/polars/py-polars/html/reference/expressions/api/polars.Expr.dt.convert_time_zone.html`

Handling duplicate values

Dealing with duplicate values is one of the common challenges we encounter when analyzing data or building data transformations. There are DataFrame/Series methods and expressions to find duplicate values, remove them, and extract only unique values.

In this recipe, we'll cover how to check and handle duplicate values in Polars.

How to do it...

Here are the steps:

1. Check the shape of the dataset:

    ```
    df.shape
    ```

    The preceding code will return the following output:

    ```
    >> (137700, 16)
    ```

2. Check the number of duplicated/unique rows at the dataset level with all the columns:

    ```
    df.is_duplicated().sum()
    ```

    The preceding code will return the following output:

    ```
    >> 0
    ```

    ```
    df.is_unique().sum()
    ```

    The preceding code will return the following output:

    ```
    >> 137700
    ```

    ```
    df.n_unique()
    ```

    The preceding code will return the following output:

    ```
    >> 137700
    ```

3. Display the number of unique values for selected columns:

    ```
    df.select(pl.all().n_unique())
    ```

    The preceding code will return the following output:

shape: (1, 16)

Data As Of	Start Date	End Date	Group	Year	...	Pneumonia Deaths	Pneumonia and COVID-19 Deaths	Influenza Deaths	Pneumonia, Influenza, or COVID-19 Deaths	Footnote
u32	u32	u32	u32	u32	...	u32	u32	u32	u32	u32
1	45	45	3	5	...	3556	2533	493	4264	2

Figure 3.10 – The number of unique values in each column

4. Display the number of unique values for a combination of columns:

```
df.n_unique(subset=['Start Date', 'End Date'])
```

The preceding code will return the following output:

```
>> 50
```

5. Here's a way to identify distinct values based on the selected subset of columns:

```
(
    df
    .unique(subset=['Start Date', 'End Date'], keep='first')

    .head())
```

The preceding code will return the following output:

Figure 3.11 – The first five rows of distinct values based on the subset of columns

6. Drop duplicated rows:

```
rows_to_keep = df.select(['Year', 'COVID-19 Deaths']).is_
unique()
rows_to_keep.sum()
```

The preceding code will return the following output:

```
>> 3940
```

```
df.filter(rows_to_keep).shape
```

The preceding code will return the following output:

```
>> (3940, 16)
```

How it works...

When using methods/expressions such as .is_duplicated() and .is_unique(), you can utilize .sum() to get the total rows since those methods return Boolean values (true evaluates to 1 and false evaluates to 0). The .n_unique() method/expression does the same job of returning the number of unique rows.

When `.is_duplicated()` or `.is_unique()` is used in conjunction with `.filter()`, you can filter out unnecessary rows from a DataFrame.

The `.unique()` method/expression gives you the unique combination of the specified columns. Also, you can specify duplicated rows to keep with the `keep` parameter.

There's more...

When you only need the approximate number of unique values, you can use the `.approx_n_unique()` method/expression. It uses the HyperLogLog++ algorithm, which estimates the unique values efficiently. If you need accurate unique values, then use the other options demonstrated previously.

```
df.select(pl.all().approx_n_unique())
```

The preceding code will return the following output:

shape: (1, 16)

Data As Of	Start Date	End Date	Group	Year	...	Pneumonia Deaths	Pneumonia and COVID-19 Deaths	Influenza Deaths	Pneumonia, Influenza, or COVID-19 Deaths	Footnote
u32	u32	u32	u32	u32	...	u32	u32	u32	u32	u32
1	45	45	3	5	...	3544	2539	491	4294	2

Figure 3.12 – Approximate unique counts

See also

Please use these resources for additional information:

- https://pola-rs.github.io/polars/py-polars/html/reference/dataframe/api/polars.DataFrame.is_duplicated.html

- https://pola-rs.github.io/polars/py-polars/html/reference/dataframe/api/polars.DataFrame.is_unique.html

- https://pola-rs.github.io/polars/py-polars/html/reference/dataframe/api/polars.DataFrame.n_unique.html

- https://pola-rs.github.io/polars/py-polars/html/reference/dataframe/api/polars.DataFrame.approx_n_unique.html

- https://docs.pola.rs/api/python/stable/reference/expressions/api/polars.approx_n_unique.html

Masking sensitive data

Security and privacy are crucial when working with data. There is some data that only certain people are allowed to see, such as your social security number, driver's license number, and medical records. This data should be treated with care and be protected appropriately.

The best scenario is that you don't store these kinds of data, or the process of masking data has been taken care of before the data gets to you. However, it's always good to know how to work with and hide this kind of information. There are some ways you can mask your data, including replacing or randomizing values, hashing, and encryption.

In this recipe, we'll cover how to mask our data by replacing values as well as hashing them.

How to do it...

Here is how to mask sensitive data:

1. Create a column called SSN (which stands for social security number).

    I.   Create a function to generate random numbers:

```
import random
def get_random_nums(num_list, length):
    random_nums = ''.join(str(n) for n in random.sample(num_
list, length))
    return random_nums
```

    II.   Generate random numbers and put those in a DataFrame:

```
fake_ssns = []
nums = [n for n in range(1, 10)]

for i in range(df.height):
    part_1 = get_random_nums(nums, 3)
    part_2 = get_random_nums(nums, 2)
    part_3 = get_random_nums(nums, 4)
    fake_ssn = f'{part_1}-{part_2}-{part_3}'
    fake_ssns.append(fake_ssn)

random.seed(10)
fake_ssns_df = pl.DataFrame({'SSN': fake_ssns})
fake_ssns_df.head()
```

The preceding code will return the following output:

shape: (5, 1)

SSN
str
"487-92-8642"
"234-73-2867"
"948-84-8635"
"942-38-1296"
"496-73-6312"

Figure 3.13 – Fake social security numbers

III.   Combine the original DataFrame and the new one that contains the SSN column:

```
df = pl.concat([df, fake_ssns_df], how='horizontal')
```

2. Replace most of SSN with other string values except for the last two digits:

```
df.select(
    ('XXX-XX-XX' + pl.col('SSN').str.slice(9, 2)).alias('SSN
Masked')
).head()
```

The preceding code will return the following output:

shape: (5, 1)

SSN Masked
str
"XXX-XX-XX16"
"XXX-XX-XX10"
"XXX-XX-XX29"
"XXX-XX-XX42"
"XXX-XX-XX75"

Figure 3.14 – Masked social security numbers

3. Use hashing to mask the social security numbers:

```
df.select(
    pl.col('SSN').hash()
).head()
```

The preceding code will return the following output:

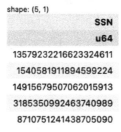

shape: (5, 1)

SSN
u64
13579232216623324611
15405819118945992244
14915679507062015913
31853509924637409899
8710751241438705090

Figure 3.15 – Hashed social security numbers

How it works...

Under the str namespace, there are a number of methods available to help with string manipulations. In this example, we used .str.slice(), which takes a subset of a string based on the specified starting index and the number of characters to take. We cover string manipulations in detail in *Chapter 6, Performing String Manipulations*.

In *step 1*, you will have seen `random.seed(10)`. It is there for reproducibility. Otherwise, the code would generate different patterns every time you run it.

There's more…

As mentioned in the introduction of this recipe, encryption is one of the ways to protect your sensitive data. There are Python libraries available to help implement encryption, such as the *cryptography* library. You can learn more about it in the documentation: `https://cryptography.io/en/latest/`.

See also

For additional information, please refer to these resources:

- `https://pola-rs.github.io/polars/py-polars/html/reference/api/polars.concat.html`

- `https://pola-rs.github.io/polars/py-polars/html/reference/expressions/api/polars.Expr.str.slice.html`

- `https://pola-rs.github.io/polars/py-polars/html/reference/expressions/api/polars.Expr.hash.html`

- `https://cryptography.io/en/latest/`

Visualizing data using Plotly

Visualizing data is essential in data analysis workflows because it simplifies complex information and highlights patterns while also improving communication and aiding in quality assessment. Additionally, data visualizations enable the detection of trends, anomalies, and relationships in data, serving as a foundational tool for exploratory data analysis.

There are many libraries available in Python to let you create visualizations, including, but not limited to, Matplotlib, Seaborn, Plotly, and Altair. Note that not all the data visualization libraries have built-in compatibility with Polars DataFrames.

In this recipe, we'll explore the data by visualizing data using the `plotly` library. It is already compatible with Polars DataFrames.

Getting ready

You need to install `plotly` for this recipe. Use the following command to install it with `pip`:

```
>>> pip install plotly
```

You'll also need the `nbformat` library to render Plotly visualizations in Jupyter Notebook. Install it with the following command in your terminal:

```
>>> pip install nbformat
```

Make sure you import the library in your Python code. We use Plotly Express in this recipe, which is a high-level interface for Plotly:

```
import plotly.express as px
```

We'll continue to use the COVID dataset we've been using throughout this chapter. Also, you might've noticed this, but the dataset is not in its cleanest format. It contains some noise that we're not interested in for the purpose of our analysis. So, we'll filter out unnecessary parts of the data before jumping into visualizing it.

The good thing is that we're aware of this data issue during our exploration and inspection of the dataset. This kind of issue is also often found in the phase of visualizing the data to explore further and as you start analyzing it.

How to do it...

We'll start by filtering the data and then dive into visualizations:

1. Import the dataset and filter out unnecessary noise:

    ```
    import polars as pl
    import plotly.express as px
    age_groups = ['0-17 years', '18-29 years', '30-39 years', '40-49
    years', '50-64 years', '65-74 years', '75-84 years', '85 years
    and over', 'All Ages']

    df = (
        pl.read_csv('../data/covid_19_deaths.csv')
        .filter(
            pl.col('Month').is_not_null(),
            pl.col('Age Group').is_in(age_groups),
        )
    )
    ```

2. Convert the data types to the correct ones, just like we did in a previous recipe, *Casting data types*:

    ```
    df = (
        df.
        with_columns(
            pl.col('Data As Of').str.strptime(pl.Date, '%m/%d/%Y'),
            pl.col('Start Date').str.strptime(pl.Date, '%m/%d/%Y'),
            pl.col('End Date').str.strptime(pl.Date, '%m/%d/%Y'),
    ```

```
            pl.col('Year').cast(pl.Int64),
            pl.col('Month').cast(pl.Int64)
        )
    )
```

3. Show COVID deaths in 2023 in the US by age group:

```
import plotly.express as px

covid_deaths_by_age = (
    df
    .filter(
        pl.col('State') == 'United States',
        pl.col('Year') == 2023,
        pl.col('Age Group') != 'All Ages',
        pl.col('Sex') == 'All Sexes'
    )
    .group_by('Age Group')
    .agg(pl.col('COVID-19 Deaths').sum())
    .sort(by='COVID-19 Deaths', descending=True)
)

fig = px.bar(
    covid_deaths_by_age,
    x='Age Group',
    y='COVID-19 Deaths',
    title='COVID Deaths 2023 by Age Group - As of 9/27/23',
    labels={'Age Group': ''}
)

fig.show()
```

The preceding code will return the following output:

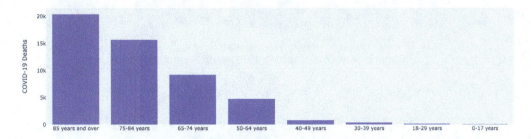

Figure 3.16 – COVID deaths in 2023 by age group

4. Display the number of COVID deaths in 2023 in the US by the top five states:

```python
covid_deaths_by_top_5_states = (
    df
    .filter(
        pl.col('State') != 'United States',
        pl.col('Year') == 2023,
        pl.col('Age Group') == 'All Ages',
        pl.col('Sex') == 'All Sexes'
    )
    .group_by('State')
    .agg(pl.col('COVID-19 Deaths').sum())
    .sort(by='COVID-19 Deaths', descending=True)
    .head()
)

fig = px.bar(
    covid_deaths_by_top_5_states,
    x='State',
    y='COVID-19 Deaths',
    title='COVID Deaths 2023 by Top 5 States - As of 9/27/23',
    labels={'State': ''}
)

fig.show()
```

The preceding code will return the following output:

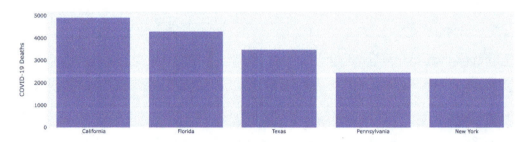

Figure 3.17 – COVID deaths in 2023 by top five states

5. Display COVID deaths in 2023 in the US by sex, with data labels:

```python
covid_deaths_by_sex = (
    df
    .filter(
```

```
            pl.col('State') == 'United States',
            pl.col('Year') == 2023,
            pl.col('Age Group') == 'All Ages',
            pl.col('Sex') != 'All Sexes'
        )
        .group_by('Sex')
        .agg(pl.col('COVID-19 Deaths').sum())
        .sort(by='COVID-19 Deaths', descending=True)
        .head()
    )

    fig = px.bar(
        covid_deaths_by_sex,
        x='Sex',
        y='COVID-19 Deaths',
        title='COVID Deaths 2023 by Sex - As of 9/27/23',
        labels={'Sex': ''},
        text_auto='.2s'
    )

    fig.update_traces(width = 0.3, textfont_size=12, textangle=0,
    textposition='inside')
    fig.show()
```

6. The preceding code will return the following output:

Figure 3.18 – COVID deaths in 2023 by sex

7. Build a scatterplot to analyze multiple variables. The following code involves a custom module, `us_state_mappings.py`. You'll need to have it in a place that makes it importable. For example, if you have it in the same folder as the notebook you're running the code from, it'll work just fine.

```
from us_state_mappings import us_state_division_dict

covid_deaths_vs_flu_deaths = (
```

```
        df
        .with_columns(
            pl.col('State').replace_strict(us_state_division_dict,
default='Others').alias('Division')
        )
        .filter(
            pl.col('State') != 'United States',
            pl.col('Age Group') != 'All Ages',
            pl.col('Sex') != 'All Sexes',
            pl.col('Year') == 2023
        )
        .group_by('State', 'Division')
        .agg(
            pl.col('COVID-19 Deaths').sum(),
            pl.col('Influenza Deaths').sum(),
            pl.col('Pneumonia Deaths').sum()
        )
)


fig = px.scatter(
    covid_deaths_vs_flu_deaths,
    x='COVID-19 Deaths',
    y='Influenza Deaths',
    color='Division',
    size='Pneumonia Deaths',
    hover_name='State',
    title='COVID-19, Influenza, and Pneumonia Deaths 2023 by US
States and Divisions'
)

fig.show()
```

The preceding code will return the following output:

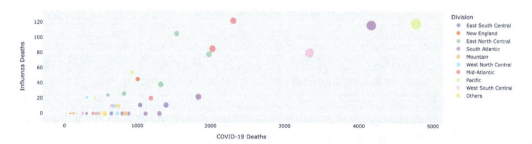

Figure 3.19 – COVID, flu, and pneumonia deaths in 2023 by states and divisions

8. Show the monthly trend by year:

```
monthly_treand_by_year = (
    df
    .filter(
        pl.col('State') == 'United States',
        pl.col('Age Group') == 'All Ages',
        pl.col('Sex') == 'All Sexes'
    )
    .group_by('Year', 'Month')
    .agg(
        pl.col('COVID-19 Deaths').sum(),
    )
    .sort(by='Month')
)

fig = px.line(
    monthly_treand_by_year,
    x='Month',
    y='COVID-19 Deaths',
    color='Year',
    title='COVID-19 Deaths Monthly Trend - United States',
    line_shape='spline'
)

fig.update_xaxes(dtick = 1)
fig.update_layout(legend_traceorder='reversed')
fig.show()
```

The preceding code will return the following output:

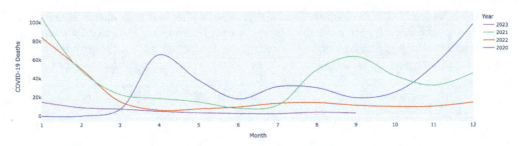

Figure 3.20 – COVID deaths in 2023 monthly trend

How it works...

If you've worked with a data visualization library in Python before, the syntax for using Plotly express should look familiar. It's not much different in that you first aggregate your data and add details if needed. You basically add the chart type to build a visualization figure, such as `px.bar()` or `px.line()`. Plotly offers many types of visualizations and they come with the flexibility to adjust parameters as well.

One advantage of using visualizations in Plotly is that they look better visually compared to the ones created in libraries such as Matplotlib. They also allow for interactivity where you can zoom in and out on a visual and hover over it for some details.

The `.group_by()` and `.agg()` methods are used to aggregate values based on certain groups you specify. We'll dive into grouping and aggregations in detail in the next chapter.

For the scatterplot shown previously, we used `.replace_strict()` to add values for a column based on the mappings from a Python dictionary. It is a useful method that you can use to simplify your transformation logic. There is also a method called `.replace()`. The difference between `.replace()` and `.replace_strict()` is that the former retains the existing data type and is intended for substituting some values in your existing column. The latter is designed for mapping some or all values from the original column with an option to set a default value and a different resulting data type. If no default value is specified, the method will raise an error if any non-null values are left unmapped. In short, use `.replace()` if you want to replace some values in your existing column and you want to keep the existing data type. Otherwise, use `.replace_strict()`. You can learn more about these methods in the Polars documentation:

- `https://docs.pola.rs/api/python/stable/reference/expressions/api/polars.Expr.replace_strict.html`

- `https://docs.pola.rs/api/python/stable/reference/expressions/api/polars.Expr.replace.html`

> **Important**
>
> Since Polars' version 0.20.3, the built-in visualization capability has been introduced. You need hvplot>=0.9.1 installed. To learn more about how to use the built-in visualizations, please refer to hvplot documentation pages: `https://hvplot.holoviz.org/`. Polars User guide on visualizations: `https://docs.pola.rs/user-guide/misc/visualization/`.

See also

Please refer to these resources to learn more about visualizations using Plotly:

- `https://plotly.com/python/bar-charts/`
- `https://plotly.com/python/line-charts/`
- `https://plotly.com/python/line-and-scatter/`

Detecting and handling outliers

Detecting and handling outliers in data analysis and science projects is essential due to their potential to distort results and undermine data quality. Outliers can significantly impact predictive models, disrupt data visualizations, and violate assumptions made in the analysis.

Addressing outliers not only enhances the robustness of the analysis but also facilitates more accurate inferences, making the insights more interpretable. This is crucial for informed decision-making. Moreover, some industries have regulatory requirements mandating the detection and handling of outliers, adding an extra layer of importance to this process.

In this recipe, we'll cover a few methods for detecting outliers as well as handling them.

Getting ready

Just like in the previous recipe, you need to install Plotly. Use the following command to install it in `pip` if you haven't already:

```
>>> pip install plotly
```

You'd also need the `nbformat` library to render Plotly visualizations in Jupyter Notebook. Install it with the following command in your terminal:

```
>>> pip install nbformat
```

Make sure you import the library in your Python code. We use Plotly express in this recipe, which is a high-level interface for Plotly:

```
import plotly.express as px
```

In this recipe, we'll be using the classic iris dataset from the Plotly library.

How to do it...

Here are the steps for detecting and handling outliers:

1. We will detect outliers using the **interquartile range (IQR)**. We're specifically looking at `sepal_width`.

    I.   Get the dataset into a DataFrame:

    ```
    import polars as pl
    import plotly
    df = pl.from_pandas(plotly.data.iris())
    df.head()
    ```

    II.   Visualize it with a boxplot:

    ```
    import plotly.express as px

    fig = px.box(df, y='sepal_width', width=500)
    fig.show()
    ```

    The preceding code will return the following output, including the tooltips on hover:

    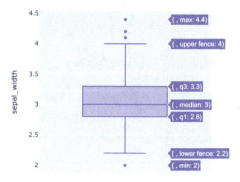

    Figure 3.21 – Boxplot for sepal_width

    If you hover over the visualization, you can see the detailed statistics of the boxplot.

2. Remove or replace outliers with the IQR.

    I.   Calculate the IQR:

    ```
    q1 = pl.col('sepal_width').quantile(0.25)
    q3 = pl.col('sepal_width').quantile(0.75)
    iqr = q3 - q1
    threshold = 1.5
    ```

```
lower_limit = q1 - iqr * threshold
upper_limit = q3 + iqr * threshold
```

II. Check the outliers. They match what we see in the preceding visual:

```
df.filter(
    (pl.col('sepal_width') < lower_limit) | (pl.col('sepal_
width') > upper_limit)
).head()
```

The preceding code will return the following output:

shape: (4, 6)

sepal_length	sepal_width	petal_length	petal_width	species	species_id
f64	f64	f64	f64	str	i64
5.7	4.4	1.5	0.4	"setosa"	1
5.2	4.1	1.5	0.1	"setosa"	1
5.5	4.2	1.4	0.2	"setosa"	1
5.0	2.0	3.5	1.0	"versicolor"	2

Figure 3.22 – The list of outliers

III. Remove the outliers:

```
is_outlier_iqr = (pl.col('sepal_width') < lower_limit) |
(pl.col('sepal_width') > upper_limit)
df_iqr_outlier_removed = (
    df
    .filter(is_outlier_iqr.not_())
)
df_iqr_outlier_removed.filter(is_outlier_iqr)
```

The preceding code will return the following output:

shape: (0, 6)

sepal_length	sepal_width	petal_length	petal_width	species	species_id
f64	f64	f64	f64	str	i64

Figure 3.23 – Confirmed the outliers have been removed from the dataset

IV. Replace the outliers with the median and check there are no outliers after replacing them:

```
df_iqr_outlier_replaced = (
    df
    .with_columns(
        pl.when(is_outlier_iqr)
```

```
            .then(pl.col('sepal_width').median())
            .otherwise(pl.col('sepal_width'))
            .alias('sepal_width')
        )
    )
df_iqr_outlier_replaced.filter(is_outlier_iqr)
```

The preceding code will return the following output:

shape: (0, 6)

sepal_length	sepal_width	petal_length	petal_width	species	species_id
f64	f64	f64	f64	str	i64

Figure 3.24 – Confirmed outliers are no longer in the dataset

3. Detect outliers with the z-score. Add a new column that calculates the z-score for every data point for the `sepal_length` column:

```
df_zscore = (
    df.with_columns(
        sepal_width_zscore=(pl.col('sepal_width') -
pl.col('sepal_width').mean()) / pl.col('sepal_width').std()
    )
)
df_zscore.head()
```

The preceding code will return the following output:

shape: (5, 7)

sepal_length	sepal_width	petal_length	petal_width	species	species_id	sepal_width_zscore
f64	f64	f64	f64	str	i64	f64
5.1	3.5	1.4	0.2	"setosa"	1	1.028611
4.9	3.0	1.4	0.2	"setosa"	1	-0.12454
4.7	3.2	1.3	0.2	"setosa"	1	0.33672
4.6	3.1	1.5	0.2	"setosa"	1	0.10609
5.0	3.6	1.4	0.2	"setosa"	1	1.259242

Figure 3.25 – Added the z-score as a column

4. Remove or replace outliers with the z-score.

    I. Remove the outliers:

```
outliers_z_score = (pl.col('sepal_width_zscore') > 3) | (pl.
col('sepal_width_zscore') < -3)
df_zscore_outliers_removed = df_zscore.filter(outliers_z_
```

```
        score.not_())
```

II.   Let's check how many points were outliers:

```
        df_zscore.filter(is_outlier_z_score)
```

The preceding code will return the following output:

shape: (1, 7)

sepal_length	sepal_width	petal_length	petal_width	species	species_id	sepal_width_zscore
f64	f64	f64	f64	str	i64	f64
5.7	4.4	1.5	0.4	"setosa"	1	3.104284

Figure 3.26 – Show outliers based on the z-score

III.   Check the outliers were removed:

```
        df_zscore_outliers_removed.filter(is_outlier_z_score)
```

The preceding code will return the following output:

shape: (0, 7)

sepal_length	sepal_width	petal_length	petal_width	species	species_id	sepal_width_zscore
f64	f64	f64	f64	str	i64	f64

Figure 3.27 – Confirmed the outliers were removed

IV.   Replace the outliers with the mean:

```
        df_zscore_outliers_replaced = (
            df_zscore
            .with_columns(
                pl.when(is_outlier_z_score)
                .then(pl.col('sepal_width').mean())
                .otherwise(pl.col('sepal_width'))
                .alias('sepal_width')
            )
        )
        df_zscore_outliers_replaced.filter(is_outlier_z_score)
```

The preceding code will return the following output:

shape: (1, 7)

sepal_length	sepal_width	petal_length	petal_width	species	species_id	sepal_width_zscore
f64	f64	f64	f64	str	i64	f64
5.7	3.054	1.5	0.4	"setosa"	1	3.104284

Figure 3.28 – Replaced outliers with the mean value

How it works...

When using the z-score to handle outliers, +3/-3 is often used as the cutoff value. The formula for calculating the z-score is *(observed value – mean) / standard deviation*.

For the IQR, you calculate the 25th (Q1) and 75th (Q3) percentiles, get the difference, and multiply by the threshold number. Then, if you're trying to calculate the lower limit, you subtract that result from Q1. If you're trying to calculate the upper limit, then you add it to Q3. A boxplot calculates all that for you.

`pl.when()` is very handy for implementing conditional logic. If you're familiar with pandas/NumPy, you can use it like you would `np.where()`. Or you can say it's like the `case` when statement in SQL. `pl.when()` is very explicit and easy to read and understand.

Also, one unique feature of Polars is that you can store your expressions in a variable and reference it in Polars' contexts for later use:

```
is_outlier_iqr = (pl.col('sepal_width') < lower_limit) | (pl.
col('sepal_width') > upper_limit)
```

In addition, Polars has a number of statistical methods for column expressions such as `.median()` and `.mean()`.

There's more...

When building a boxplot in Plotly, you can choose the quartile algorithm. Your options are `linear`, `inclusive`, and `exclusive`. For additional information, please refer to this Plotly documentation: `https://plotly.com/python/box-plots/`.

See also

You can find additional information using these resources:

- `https://en.wikipedia.org/wiki/Interquartile_range`
- `https://en.wikipedia.org/wiki/Standard_score`
- `https://pola-rs.github.io/polars/py-polars/html/reference/expressions/api/polars.when.html`
- `https://pola-rs.github.io/polars/py-polars/html/reference/expressions/api/polars.quantile.html`
- `https://plotly.com/python/box-plots/`

4

Data Transformation Techniques

In this chapter, we will look at how aggregations, window functions, and **User-Defined Functions (UDFs)** are essential tools in data analysis, data science, and data engineering workflows. We'll also cover how we can use SQL in Python Polars.

We will understand how aggregations involve combining and summarizing data to gain insights. They are commonly used in data analysis to perform operations such as sum, average, count, or maximum on a dataset. They help summarize your data and compute the necessary parts to further your data transformations.

We will also understand how window functions, on the other hand, allow you to perform calculations across a specific window or subset of data within a dataset. They are valuable in data analysis for tasks such as ranking and identifying trends within a partition of data.

Furthermore, we will learn about UDFs that provide flexibility by allowing you to define custom functions to process and transform data. There is a possibility that the types of transformation you're trying to implement are hard to accomplish in Python Polars. In that case, UDFs help provide the ability to tailor data processing to the unique needs of a project by creating custom functions to apply to the dataset.

Overall, these three tools play a central role in manipulating and extracting insights from data in a variety of data-related workflows.

In this chapter, we're going to cover the following recipes:

- Exploring basic aggregations
- Using group by aggregations
- Aggregating values across multiple columns
- Computing with window functions

- Applying UDFs
- Using SQL for data transformations

Technical requirements

You can download the datasets and code in the GitHub repository:

- Data: `https://github.com/PacktPublishing/Polars-Cookbook/tree/main/data`
- Code: `https://github.com/PacktPublishing/Polars-Cookbook/tree/main/Chapter04`

It is assumed that you have installed the Polars library in your Python environment:

```
>>> pip install polars
```

It is also assumed that you imported it in your code:

```
import polars as pl
```

In this chapter, we'll be using the Contoso dataset (`https://github.com/PacktPublishing/Polars-Cookbook/blob/main/data/contoso_sales.csv`), which contains sales data. It also includes attributes such as products and customers and we'll look at aggregating sales by these dimensions using various aggregation techniques.

Exploring basic aggregations

Simple aggregations such as sum, mean, count, max, min, and so on are fundamental techniques to analyze and prepare your data. They provide a quick and efficient way to summarize and gain insights from large datasets. These operations also allow you to distill complex information into manageable, interpretable results.

In this recipe, we'll cover how to use simple aggregations at the DataFrame and Series levels as well as in Polars' expressions.

How to do it...

Here are the steps for exploring basic aggregations:

1. Read the dataset into a DataFrame:

    ```
    df = pl.read_csv('../data/contoso_sales.csv', try_parse_
    dates=True)
    ```

2. Calculate aggregations at the DataFrame level, selecting only numeric columns:

```
from polars import selectors as cs
(
    df
    .select(cs.numeric())
    .sum()
)
```

The preceding code will return the following output:

shape: (1, 8)

Order Number	Line Number	Customer Age	Quantity	Unit Price	Net Price	Unit Cost	Exchange Rate
i64	i64	i64	i64	f64	f64	f64	f64
4466019052	16195	725757	43517	4.1785e6	3.9286e6	1.7356e6	14124.4597

Figure 4.1 – The result of sum on the entire DataFrame

3. Get the sum of sales at the Series level:

```
s = df.select('Quantity').to_series()
s.sum()
```

The preceding code will return the following output:

```
>> 43517
```

4. This is how you can calculate aggregations with expressions:

```
df.select(pl.col('Quantity').sum())
```

The preceding code will return the following output:

shape: (1, 1)

Quantity
i64
43517

Figure 4.2 – The result of sum on a single column

5. There are a few methods that help extract the first or last element:

```
df.select(
    pl.col('Customer Name').first().alias('Cust Name First'),
    pl.col('Customer Name').last().alias('Cust Name Last')
)
```

The preceding code will return the following output:

shape: (1, 2)

Cust Name First	Cust Name Last
str	str
"Eric Kennedy"	"Billy Ratliff"

Figure 4.3 – The first and last value of the Customer Name column

6. You can get various aggregations over your entire DataFrame with `.describe()`:

```
df.select(cs.numeric()).describe()
```

The preceding code will return the following output:

shape: (9, 9)

statistic	Order Number	Line Number	Customer Age	Quantity	Unit Price	Net Price	Unit Cost	Exchange Rate
str	f64	f64	f64	f64	f64	f64	f64	f64
"count"	13915.0	13915.0	13915.0	13915.0	13915.0	13915.0	13915.0	13915.0
"null_count"	0.0	0.0	0.0	0.0	0.0	0.0	0.0	0.0
"mean"	320949.985771	1.163852	52.15645	3.127345	300.28425	282.32739	124.731364	1.015053
"std"	28431.79136	1.361349	19.133881	2.233597	405.538975	381.738847	147.944094	0.171927
"min"	269500.0	0.0	19.0	1.0	0.95	0.8265	0.48	0.7015
"25%"	295902.0	0.0	36.0	1.0	46.99	43.4	21.92	0.8965
"50%"	319806.0	1.0	52.0	2.0	207.987	194.91	86.68	1.0
"75%"	345106.0	2.0	68.0	4.0	361.2	336.0	160.93	1.0
"max"	371503.0	6.0	85.0	10.0	3748.5	3748.5	1241.955	1.5373

Figure 4.4 – The summarized information of the DataFrame

How it works...

The simple aggregations we just covered are available in DataFrame, LazyFrame, Series, and expressions. The syntax is the same across objects, so you basically add an aggregation such as `.sum()` to your DataFrame, Series, or expression.

The `.select()` method returns a DataFrame. You'd need to use `.to_series()` to convert it from a single-column DataFrame to a Series.

Selectors in Polars allow you to select columns in various ways such as data types.

Please refer to the *See also* section for additional information about basic aggregations in Polars.

There's more...

You can aggregate values with conditions. In that case, use parentheses to enclose the logic:

```
df.select(
    (pl.col('Quantity') >= 4).sum()
)
```

The preceding code will return the following output:

shape: (1, 1)

Quantity
u32
4423

Figure 4.5 – An aggregation with a condition

Use `.filter()` or `.where()` to aggregate with conditions based on other columns you're not trying to aggregate:

```
df.select(
    pl.col('Quantity').filter(pl.col('Store Name')=='Online store').
sum()
)
```

The preceding code will return the following output:

shape: (1, 1)

Quantity
i64
25017

Figure 4.6 – An aggregation with a condition on another column

See also

If you're interested in learning more about basic aggregations, please refer to the following resources:

- https://pola-rs.github.io/polars/py-polars/html/reference/dataframe/aggregation.html

- https://pola-rs.github.io/polars/py-polars/html/reference/series/aggregation.html

- https://pola-rs.github.io/polars/py-polars/html/reference/lazyframe/aggregation.html

- https://pola-rs.github.io/polars/py-polars/html/reference/expressions/aggregation.html

Using group by aggregations

Group by aggregations are essential in data analysis and involve dividing a dataset into distinct groups based on categorical values and, subsequently, applying aggregate functions to each group.

This technique is particularly useful for obtaining summary statistics and insights within subsets of the data. This not only simplifies the analysis process but also provides a more nuanced understanding of the underlying patterns and trends within the data.

In this recipe, we'll cover how to group your DataFrame and LazyFrame and apply aggregations to each group.

Getting ready

Make sure to read the Contoso sales dataset:

```
df = pl.read_csv('../data/contoso_sales.csv', try_parse_dates=True)
```

How to do it...

Here's how you can use group by aggregations:

1. Group your DataFrame by a column called `Brand`:

   ```
   df.group_by('Brand')
   ```

   The preceding code will return the following output:

   ```
   >> <polars.dataframe.group_by.GroupBy at 0x106c07e50>
   ```

2. Add an aggregation and show the first five rows:

   ```
   (
       df
       .group_by('Brand')
       .agg(pl.col('Quantity').sum().alias('Sum of Quantity'))
       .head()
   )
   ```

   The preceding code will return the following output. Note that you may see a different order of rows:

shape: (5, 2)

Brand	Sum of Quantity
str	i64
"A. Datum"	555
"Northwind Trad…	638
"Fabrikam"	1516
"Adventure Work…	4616
"Litware "	161

Figure 4.7 – A simple group by aggregation

3. Add multiple aggregations. Sort the DataFrame by the average unit price in descending order:

```
(
    df
    .group_by('Brand')
    .agg(
        pl.col('Quantity').sum().alias('Sum of Quantity'),
        pl.col('Unit Price').mean().alias('Average Unit Price'),
    )
    .sort('Average Unit Price', descending=True)
    .head()
)
```

The preceding code will return the following output:

shape: (5, 3)

Brand	Sum of Quantity	Average Unit Price
str	i64	f64
"Fabrikam "	332	795.820192
"Contoso "	977	715.378268
"Adventure Work…	4616	620.330242
"Litware "	161	603.528818
"Fabrikam"	1516	557.956117

Figure 4.8 – Multiple group by aggregations with sorting

4. You can apply more complex expressions in aggregations. Let's calculate the average in two different ways and a few other aggregations:

```
(
    df
    .group_by('Brand')
    .agg(
```

```
        pl.col('Unit Price').mean().round(2).alias('Average Unit
Price'),
        (pl.col('Unit Price').sum() / pl.count()).round(2).
alias('Average Unit Price 2'),
        pl.col('Customer Name').first(),
        pl.col('Category').last()
    )

        .sort('Average Unit Price', descending=True)
    .sort('Brand')
    .head()
)
```

The preceding code will return the following output:

shape: (5, 5)

Brand	Average Unit Price	Average Unit Price 2	Customer Name	Category
str	f64	f64	str	str
"A. Datum"	280.1	280.1	"Blažena Salabo...	"Cameras and ca...
"Adventure Work...	620.33	620.33	"Molly Walters"	"Home Appliance...
"Adventure Work...	166.9	166.9	"James Steinfel...	"TV and Video"
"Contoso"	150.86	150.86	"Eric Kennedy"	"Cell phones"
"Contoso "	715.38	715.38	"Chiquita Boyd"	"Home Appliance...

Figure 4.9 – More complex group by aggregations

5. You can accomplish the same thing with a LazyFrame as well:

```
(
    pl.scan_csv('../data/contoso_sales.csv', try_parse_
dates=True)
    .group_by('Brand')
    .agg(
        pl.col('Unit Price').mean().round(2).alias('Average Unit
Price'),
        (pl.col('Unit Price').sum() / pl.count()).round(2).
alias('Average Unit Price 2'),
        pl.col('Customer Name').first(),
        pl.col('Category').last()
    )
    .sort('Average Unit Price', descending=True)
    .sort('Brand')
    .collect()
    .head()
)
```

The preceding code will return the following output:

shape: (5, 5)

Brand	Average Unit Price	Average Unit Price 2	Customer Name	Category
str	f64	f64	str	str
"A. Datum"	280.1	280.1	"Blažena Salabo...	"Cameras and ca...
"Adventure Work...	620.33	620.33	"Molly Walters"	"Home Appliance...
"Adventure Work...	166.9	166.9	"James Steinfel...	"TV and Video"
"Contoso"	150.86	150.86	"Eric Kennedy"	"Cell phones"
"Contoso "	715.38	715.38	"Chiquita Boyd"	"Home Appliance...

Figure 4.10 – Group by aggregations with a LazyFrame

How it works...

Group by aggregation is one of the Polars contexts in which expressions work and come in handy just like selection and filtering. You can use a column expression or column name inside `group_by()`. You can also chain your expressions to build more complex business logic within the group by aggregation methods.

> **Tip**
>
> Notice some values in columns such as `Brand` are cut off when displaying your DataFrame. You can set the maximum string length to display by adjusting the value for `pl.Config.set_fmt_str_lengths` for printing in a console or terminal. When displaying in Jupyter Notebook, though, you need to adjust the environmental variable (as of the time of writing the book). Refer to the following examples to understand what I mean. Find more information about adjusting the maximum string length to display here: `https://pola-rs.github.io/polars/py-polars/html/reference/api/polars.Config.set_fmt_str_lengths.html#polars.Config.set_fmt_str_lengths`.

Setting `pl.Config.set_fmt_str_lengths` helps you output your DataFrame in a more readable matter with `print()`:

```
pl.Config.set_fmt_str_lengths = 50
print(df.select('Brand').unique().head(10))
```

The preceding code will return the following output:

Figure 4.11 – An example output with complete string values

But that setting doesn't work when displaying your DataFrame in Jupyter Notebook, as you will see:

```
df.select('Brand').unique().head(10)
```

The preceding code will return the following output:

Figure 4.12 – The same code doesn't work in Jupyter Notebook

You need to manually set the environment variable that Polars uses for its configurations:

```
import os
os.environ['POLARS_FMT_STR_LEN'] = str(50)


df.select('Brand').unique().head(10)
```

The preceding code will return the following output:

shape: (10, 1)

Brand
str
"Contoso"
"Wide World Importers"
"Adventure Works "
"Tailspin Toys"
"Fabrikam"
"Litware "
"Adventure Works"
"Litware"
"Southridge Video"
"A. Datum"

Figure 4.13 – The output after adjusting the environment variable

There's more...

There are a few things worth mentioning.

Firstly, group_by objects are iterable. You can iterate through a group by object and return a DataFrame for each group:

```
for name, data in df.group_by(['Brand']):
    print(name[0], type(data))
```

The preceding code will return the following output:

```
Proseware <class 'polars.dataframe.frame.DataFrame'>
Litware  <class 'polars.dataframe.frame.DataFrame'>
A. Datum <class 'polars.dataframe.frame.DataFrame'>
Tailspin Toys <class 'polars.dataframe.frame.DataFrame'>
Contoso  <class 'polars.dataframe.frame.DataFrame'>
Fabrikam <class 'polars.dataframe.frame.DataFrame'>
Contoso <class 'polars.dataframe.frame.DataFrame'>
Adventure Works  <class 'polars.dataframe.frame.DataFrame'>
Adventure Works <class 'polars.dataframe.frame.DataFrame'>
The Phone Company <class 'polars.dataframe.frame.DataFrame'>
Northwind Traders <class 'polars.dataframe.frame.DataFrame'>
Wide World Importers <class 'polars.dataframe.frame.DataFrame'>
Litware <class 'polars.dataframe.frame.DataFrame'>
Fabrikam   <class 'polars.dataframe.frame.DataFrame'>
Southridge Video <class 'polars.dataframe.frame.DataFrame'>
```

Figure 4.14 – Iterating through a group by object

Secondly, if you use the combination of `.group_by()` and `.agg()` without any aggregation function, and only reference your column, it transforms the list of values for each group into a list:

```
(
    df
    .group_by('Brand')
    .agg(pl.col('Quantity'))
    .head()
)
```

The preceding code will return the following output. Note that you may see a different order of rows:

shape: (5, 2)

Brand	Quantity
str	list[i64]
"Litware "	[2, 1, ... 3]
"Adventure Work...	[5, 7, ... 2]
"Contoso"	[7, 1, ... 3]
"Fabrikam"	[2, 1, ... 2]
"Adventure Work...	[2, 3, ... 6]

Figure 4.15 – A list column created with a group by operation

We'll cover the list data type and how to work with it in *Chapter 7, Working with List and Array Operations*.

Thirdly, you can use a parameter in `.group_by()`. You have a parameter called `maintain_order` that, if you set to `true`, keeps the order of the groups consistent with your original data:

```
(
    df
    .group_by('Brand', maintain_order=True)
    .agg(pl.col('Quantity'))
    .head()
)
```

The preceding code will return the following output:

shape: (5, 2)

Brand	Quantity
str	list[i64]
"Contoso"	[7, 1, … 3]
"Wide World Imp…	[2, 8, … 2]
"Northwind Trad…	[2, 3, … 2]
"Adventure Work…	[2, 3, … 6]
"Adventure Work…	[5, 7, … 2]

Figure 4.16 – A DataFrame after a group by with the original order maintained

> **Note**
>
> This is slower than the `group_by` operation without setting the `maintain_order` parameter to `true`. And it possibly blocks the code running with streaming mode as well.

Lastly, you can reference column names as strings or use expressions inside `.group_by()`:

```
(
    df
    .group_by(
        pl.col('Brand'),
        'Customer Country',
        pl.col('Order Date').dt.year().alias('Order Year')
    )
    .agg(pl.col('Unit Price').mean())
    .head()
)
```

The preceding code will return the following output:

shape: (5, 4)

Brand	Customer Country	Order Year	Unit Price
str	str	i32	f64
"Contoso"	"United Kingdom...	2018	149.204372
"Contoso"	"Germany"	2018	156.208612
"Contoso"	"Italy"	2017	217.8535
"Wide World Imp...	"Netherlands"	2018	631.144348
"Wide World Imp...	"United States"	2018	513.037234

Figure 4.17 – A DataFrame with various ways to reference columns

> Tip
>
> .dt.year() is one of the temporal expressions that extract the year from a date column. .dt is what gives you access to many time-related expressions.

See also

Please refer to these resources to learn more about basic aggregations:

- https://pola-rs.github.io/polars/user-guide/expressions/aggregation/#basic-aggregations

- https://pola-rs.github.io/polars/py-polars/html/reference/dataframe/api/polars.DataFrame.group_by.html

- https://pola-rs.github.io/polars/py-polars/html/reference/api/polars.Config.set_fmt_str_lengths.html#polars.Config.set_fmt_str_lengths

- https://pola-rs.github.io/polars/py-polars/html/reference/expressions/temporal.html

Aggregating values across multiple columns

Although aggregation typically means aggregating values in a column, there are cases where you might want to aggregate values horizontally or across multiple columns. And there are methods available in Python Polars that allow you to do that.

In this recipe, we'll cover multiple ways to aggregate values across multiple columns.

Getting ready

In this recipe, we'll use a Pokémon dataset:

```
df = pl.read_csv('../data/pokemon.csv')
df.head()
```

The preceding code will return the following output:

shape: (5, 13)

#	Name	Type 1	Type 2	Total	HP	Attack	Defense	Sp. Atk	Sp. Def	Speed	Generation	Legendary
i64	str	str	str	i64	i64	i64	i64	i64	i64	i64	i64	bool
1	"Bulbasaur"	"Grass"	"Poison"	318	45	49	49	65	65	45	1	false
2	"Ivysaur"	"Grass"	"Poison"	405	60	62	63	80	80	60	1	false
3	"Venusaur"	"Grass"	"Poison"	525	80	82	83	100	100	80	1	false
3	"VenusaurMega Venusaur"	"Grass"	"Poison"	625	80	100	123	122	120	80	1	false
4	"Charmander"	"Fire"	null	309	39	52	43	60	50	65	1	false

Figure 4.18 – The first five rows of the Pokémon dataset

How to do it...

Here are ways to aggregate values across columns:

1. Calculate the sum of all the stats of each Pokémon:

```
(
    df
    .select('HP', 'Attack', 'Defense', 'Sp. Atk', 'Sp. Def',
'Speed')
    .sum_horizontal().alias('Total 2')
    .head()
)
```

The preceding code will return the following output:

shape: (5,)

Total 2
i64
318
405
525
625
309

Figure 4.19 – A column calculated with a horizontal aggregation

If you look at the Total column in your dataset, these numbers match.

2. Calculate the total again, but this time, we'll use an expression to add the result as a new column:

```
(
    df
    .with_columns(
        pl.sum_horizontal('HP', 'Attack', 'Defense', 'Sp. Atk',
'Sp. Def', 'Speed').alias('Total 2')
    )
    .head()
)
```

3. The preceding code will return the following output:

shape: (5, 14)

#	Name	Type 1	Type 2	Total	HP	Attack	Defense	Sp. Atk	Sp. Def	Speed	Generation	Legendary	Total 2
i64	str	str	str	i64	i64	i64	i64	i64	i64	i64	i64	bool	i64
1	"Bulbasaur"	"Grass"	"Poison"	318	45	49	49	65	65	45	1	false	318
2	"Ivysaur"	"Grass"	"Poison"	405	60	62	63	80	80	60	1	false	405
3	"Venusaur"	"Grass"	"Poison"	525	80	82	83	100	100	80	1	false	525
3	"VenusaurMega Venusaur"	"Grass"	"Poison"	625	80	100	123	122	120	80	1	false	625
4	"Charmander"	"Fire"	null	309	39	52	43	60	50	65	1	false	309

Figure 4.20 – A new column added that aggregates values across columns

4. Alternatively, you can use `pl.concat_list()` and apply `.sum()` over each list:

```
(
    df
    .with_columns(
        pl.concat_list('HP', 'Attack', 'Defense', 'Sp. Atk',
'Sp. Def', 'Speed').list.sum().alias('Total 2')
    )
    .head()
)
```

We'll cover list operations in *Chapter 7, Working with List and Array Operations*.

5. Use `pl.reduce()` for calculating the same, `Total 2`:

```
cols = ['HP', 'Attack', 'Defense', 'Sp. Atk', 'Sp. Def',
'Speed']
(
    df
    .with_columns(
        pl.reduce(
            function=lambda acc, col: acc + col,
            exprs=pl.col(cols)
        )
        .alias('Total 2')
    )
```

```
        .head()
)
```

The preceding code will return the following output:

shape: (5, 14)

#	Name	Type 1	Type 2	Total	HP	Attack	Defense	Sp. Atk	Sp. Def	Speed	Generation	Legendary	Total 2
i64	str	str	str	i64	i64	i64	i64	i64	i64	i64	i64	bool	i64
1	"Bulbasaur"	"Grass"	"Poison"	318	45	49	49	65	65	45	1	false	318
2	"Ivysaur"	"Grass"	"Poison"	405	60	62	63	80	80	60	1	false	405
3	"Venusaur"	"Grass"	"Poison"	525	80	82	83	100	100	80	1	false	525
3	"VenusaurMega Venusaur"	"Grass"	"Poison"	625	80	100	123	122	120	80	1	false	625
4	"Charmander"	"Fire"	null	309	39	52	43	60	50	65	1	false	309

Figure 4.21 – A horizontal aggregation with the reduce function

6. Use `pl.fold()` and add `100` to the output:

```
(
    df
    .with_columns(
        pl.fold(
            acc=pl.lit(100),
            function=lambda acc, col: acc + col, exprs=pl.
col(cols)
        )
        .alias('Total 2')
    )
    .head()
)
```

The preceding code will return the following output:

shape: (5, 14)

#	Name	Type 1	Type 2	Total	HP	Attack	Defense	Sp. Atk	Sp. Def	Speed	Generation	Legendary	Total 2
i64	str	str	str	i64	i64	i64	i64	i64	i64	i64	i64	bool	i64
1	"Bulbasaur"	"Grass"	"Poison"	318	45	49	49	65	65	45	1	false	418
2	"Ivysaur"	"Grass"	"Poison"	405	60	62	63	80	80	60	1	false	505
3	"Venusaur"	"Grass"	"Poison"	525	80	82	83	100	100	80	1	false	625
3	"VenusaurMega Venusaur"	"Grass"	"Poison"	625	80	100	123	122	120	80	1	false	725
4	"Charmander"	"Fire"	null	309	39	52	43	60	50	65	1	false	409

Figure 4.22 – A horizontal aggregation with the fold function

How it works...

Both `pl.fold()` and `pl.reduce()` can be used for both basic and complex horizontal transformation logic. The difference between `pl.fold()` and `pl.reduce()` is that the former allows you to define the initial value of the accumulator, set at the beginning of the fold, whereas that initial value is 0 for the latter.

To better understand the mechanics of these functions, here's a diagram that depicts how it works in the case of calculating the sum:

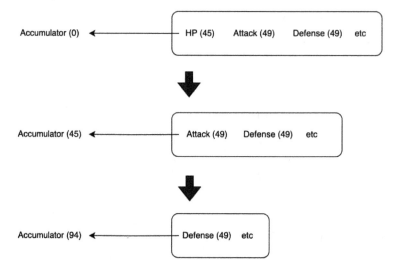

Figure 4.23 – The mechanics of pl.reduce() and pl.fold()

> **Important**
>
> When you use either pl.reduce() or pl.fold() for horizontal aggregations, if there is a null value in a row, then the resulting output becomes null for that entire row. However, if you use sum_horizontal(), null values are skipped or ignored just like any other aggregations.

There's more...

pl.fold() can not only be used for horizontal aggregations but also for conditionals or filtering rows on multiple columns:

```
(
    df
    .filter(
        pl.fold(
            acc=pl.lit(True),
            function=lambda acc, col: acc & col,
            exprs=pl.col(cols) > 80
        )
    )
    .head()
)
```

You can do the same filtering using `pl.all_horizontal()`. This is probably more descriptive and expressive than the preceding example:

```
(
    df
    .filter(
        pl.all_horizontal(pl.col(cols) > 80)
    )
    .head()
)
```

Both result in the following output:

shape: (4, 13)

#	Name	Type 1	Type 2	Total	HP	Attack	Defense	Sp. Atk	Sp. Def	Speed	Generation	Legendary
i64	str	str	str	i64	i64	i64	i64	i64	i64	i64	i64	bool
144	"Articuno"	"Ice"	"Flying"	580	90	85	100	95	125	85	1	true
145	"Zapdos"	"Electric"	"Flying"	580	90	90	85	125	90	100	1	true
146	"Moltres"	"Fire"	"Flying"	580	90	100	90	125	85	90	1	true
150	"Mewtwo"	"Psychic"	null	680	106	110	90	154	90	130	1	true

Figure 4.24 – The filtered DataFrame on multiple columns

Another thing that `pl.fold()` can do is string concatenation. However, it will be slow because it needs to cache intermediate results. In that case, you should use `pl.concat_str()` instead:

```
str_cols = ['Name', 'Type 1', 'Type 2']
str_combined_expr = pl.fold(acc=pl.lit(''), function=lambda acc, col:
acc + col, exprs=str_cols).alias('Str Combined')
str_cols.append(str_combined_expr)
df.select(str_cols).head()
```

The preceding code will return the following output:

shape: (5, 4)

Name	Type 1	Type 2	Str Combined
str	str	str	str
"Bulbasaur"	"Grass"	"Poison"	"BulbasaurGrassPoison"
"Ivysaur"	"Grass"	"Poison"	"IvysaurGrassPoison"
"Venusaur"	"Grass"	"Poison"	"VenusaurGrassPoison"
"VenusaurMega Venusaur"	"Grass"	"Poison"	"VenusaurMega VenusaurGrassPoison"
"Charmander"	"Fire"	null	null

Figure 4.25 – The DataFrame with concatenated strings

Here's the example using `pl.concat_str()`. Note that you need to redefine the `str_cols` variable for the code to work properly:

```
str_cols = ['Name', 'Type 1', 'Type 2']
df.select(pl.concat_str(str_cols)).head()
```

The preceding code will return the following output:

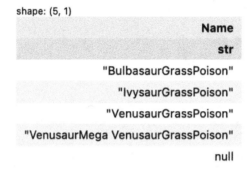

shape: (5, 1)

Name
str
"BulbasaurGrassPoison"
"IvysaurGrassPoison"
"VenusaurGrassPoison"
"VenusaurMega VenusaurGrassPoison"
null

Figure 4.26 – The concatenated strings with pl.concat_str

We'll cover string manipulations more in detail in *Chapter 6, Performing String Manipulations*.

See also

To learn more about aggregating data across columns, please refer to these additional resources:

- `https://pola-rs.github.io/polars/py-polars/html/reference/expressions/api/polars.sum_horizontal.html`

- `https://pola-rs.github.io/polars/user-guide/expressions/folds/`

- `https://pola-rs.github.io/polars/py-polars/html/reference/expressions/api/polars.fold.html`

- `https://pola-rs.github.io/polars/py-polars/html/reference/expressions/api/polars.reduce.html`

- `https://stuffbyyuki.com/aggregations-over-multiple-columns-in-polars/`

- `https://pola-rs.github.io/polars/py-polars/html/reference/expressions/api/polars.all_horizontal.html`

- `https://pola-rs.github.io/polars/py-polars/html/reference/expressions/api/polars.concat_str.html`

Computing with window functions

Window functions are a great feature in Polars. You may have heard of them or use them frequently in SQL, Spark, and so on. What they help with is the aggregations over groups. A window function helps calculate values over specific groups regardless of the granularity of your dataset. If just doing a group by operation, the resulting DataFrame is changed to the length of that group; however, with a window function, you retain the size or height of your original dataset.

In this recipe, we'll cover how to aggregate and rank rows over groups as well as sorting over groups.

Getting ready

We'll be using the Contoso dataset in this recipe. Make sure to read it in a DataFrame and create a new column beforehand:

```
df = pl.read_csv('../data/contoso_sales.csv', try_parse_dates=True)
df = df.with_columns(
    (pl.col('Quantity') * pl.col('Net Price')).round(2).alias('Sales
Amount')
)
```

Also, change the settings so that the output table shows the full string length in Jupyter Notebook:

```
import os
os.environ['POLARS_FMT_STR_LEN'] = str(50)
```

If you're just working in a Python file, you can simply use the following line of code:

```
pl.Config.set_fmt_str_lengths = 50
```

How to do it...

Here are ways to calculate window functions:

1. Calculate the sum of the sales amount over the `Category` column:

    ```
    sales_by_cat = df.select(
        'Category',
        'Subcategory',
        pl.col('Sales Amount').sum().over('Category').alias('Sales
    Amt per Cat')
    )
    sales_by_cat.head()
    ```

    The preceding code will return the following output:

shape: (5, 3)

Category	Subcategory	Sales Amt per Cat
str	str	f64
"Audio"	"MP4&MP3"	238356.0
"Audio"	"MP4&MP3"	238356.0
"Audio"	"MP4&MP3"	238356.0
"Audio"	"MP4&MP3"	238356.0
"Audio"	"MP4&MP3"	238356.0

Figure 4.27 – A window function on the Category column

Notice that the sales amount is the same for the `Audio` category, but we just looked at the first five rows. Let's check that `.over()` is working as expected by getting the unique combinations of category and subcategory for the `Audio` category:

```
sales_by_cat.filter(pl.col('Category')=='Audio').unique().head()
```

The preceding code will return the following output:

shape: (3, 3)

Category	Subcategory	Sales Amt per Cat
str	str	f64
"Audio"	"Bluetooth Headphones"	238356.0
"Audio"	"Recording Pen"	238356.0
"Audio"	"MP4&MP3"	238356.0

Figure 4.28 – The DataFrame with a new column showing sum of sales per category

You can see that the sales amount is the same across any subcategory for `Audio`.

Also, see that the original number of rows is kept:

```
df.shape, sales_by_cat.shape
```

The preceding code will return the following output:

```
>> ((13915, 21), (13915, 3))
```

2. Calculate the average sales amount over two columns, `Category` and `Brand`, and only look at the `Computers` category and the `Contoso` brand:

```
(
    df
    .select(
        'Category',
        'Brand',
        'Subcategory',
```

```
            pl.col('Sales Amount').mean().over('Category', 'Brand').
    alias('Avg Sales per Cat and Brand')
        )
        .filter((pl.col('Category')=='Computers'))
        .unique()
        .sort('Brand')
        .head(10)
    )
```

The preceding code will return the following output:

shape: (10, 4)

Category	Brand	Subcategory	Avg Sales per Cat and Brand
str	str	str	f64
"Computers"	"Adventure Works"	"Desktops"	1797.371846
"Computers"	"Adventure Works"	"Monitors"	1797.371846
"Computers"	"Adventure Works"	"Laptops"	1797.371846
"Computers"	"Contoso"	"Projectors & Screens"	689.986652
"Computers"	"Contoso"	"Computers Accessories"	689.986652
"Computers"	"Fabrikam"	"Laptops"	1982.066063
"Computers"	"Proseware"	"Laptops"	1095.305012
"Computers"	"Proseware"	"Printers, Scanners & Fax"	1095.305012
"Computers"	"Proseware"	"Monitors"	1095.305012
"Computers"	"Proseware"	"Projectors & Screens"	1095.305012

Figure 4.29 – A window function with other operations

See that the average sales are unique per category per brand.

3. Use expressions instead of just column references in .over(). The following calculates the average sales amount per category per customer birth year:

```
from datetime import date

curr_yr = date.today().year
cust_birth_yr = curr_yr - pl.col('Customer Age')

(
    df
    .select(
        'Category',
        'Brand',
```

```
        'Customer Age',
        pl.col('Sales Amount').mean().over('Category', cust_
birth_yr).alias('Avg Sales per Cat')
    )
    .filter(pl.col('Category')=='Computers')
    .unique()
    .sort('Customer Age')
    .head(10)
)
```

The preceding code will return the following output:

shape: (10, 4)

Category	Brand	Customer Age	Avg Sales per Cat
str	str	i64	f64
"Computers"	"Wide World Importers"	19	1665.993509
"Computers"	"Adventure Works"	19	1665.993509
"Computers"	"Contoso"	19	1665.993509
"Computers"	"Fabrikam"	19	1665.993509
"Computers"	"Southridge Video"	19	1665.993509
"Computers"	"Proseware"	19	1665.993509
"Computers"	"Proseware"	20	2094.541563
"Computers"	"Southridge Video"	20	2094.541563
"Computers"	"Wide World Importers"	20	2094.541563
"Computers"	"Contoso"	20	2094.541563

Figure 4.30 – Expressions inside a window function

Now, you could add `Customer Birth Year` as a column and just reference that in `.over()`. That works fine, too. My point is that being able to add expressions in `.over()` gives you a lot more flexibility in what you can do. And yes, the customer's birth year is practically the same as the customer's age...

You could also add an additional simple expression like this:

```
pl.col('Sales Amount').mean().over('Category', cust_birth_yr +
10).alias('Avg Sales per Cat')
```

4. Add rankings based on the maximum sales amount for `Category`:

```
(
    df
    .group_by('Category')
    .agg(pl.col('Sales Amount').max().alias('Max Sales Amt'))
    .with_columns(
        pl.col('Max Sales Amt').rank(descending=True)
```

```
    .alias('Rank')
    )
    .sort('Rank')
)
```

The preceding code will return the following output:

shape: (8, 3)

Category	Max Sales Amt	Rank
str	f64	f64
"TV and Video"	28999.9	1.0
"Home Appliances"	28479.91	2.0
"Computers"	19992.0	3.0
"Cameras and camcorders "	10810.8	4.0
"Cell phones"	5183.2	5.0
"Music, Movies and Audio Books"	3041.88	6.0
"Audio"	2871.2	7.0
"Games and Toys"	2813.16	8.0

Figure 4.31 – Added rank column based on max sales per category

5. Get the rankings over a specific group (Subcategory) for each category:

```
(
    df
    .group_by('Category', 'Subcategory')
    .agg(pl.col('Sales Amount').max().round().cast(pl.Int64.
alias('Max Sales Amt'))
    .with_columns(
        pl.col('Max Sales Amt').rank(descending=True).
over('Category').cast(pl.Int64).alias('Rank')
    )
    .filter(pl.col('Category').is_in(['Audio', 'Computers']))
    .sort(['Category', 'Rank'])
)
```

The preceding code will return the following output:

shape: (9, 4)

Category	Subcategory	Max Sales Amt	Rank
str	str	i64	i64
"Audio"	"Recording Pen"	2871	1
"Audio"	"Bluetooth Headphones"	2250	2
"Audio"	"MP4&MP3"	2095	3
"Computers"	"Projectors & Screens"	19992	1
"Computers"	"Laptops"	19485	2
"Computers"	"Desktops"	14535	3
"Computers"	"Monitors"	11425	4
"Computers"	"Printers, Scanners & Fax"	2508	5
"Computers"	"Computers Accessories"	2424	6

Figure 4.32 – Added rank per subcategory within each category

How it works...

If you're familiar with window functions in SQL, then understanding `.over()` is easy because it works pretty much the same way. It essentially defines the partitions and value order in which you apply your operations. As you saw in the previous examples, the syntax is very logical and clear.

The `mapping_strategy` parameter in `.over()` gives you options for what you want your resulting output to look like. By default, it's set to `group_to_rows`, which assigns the aggregations back to the original structure of your DataFrame (typical window functions). The other options are `join` and `explode`. Refer to the examples in the *There's more...* section for details on how to use them.

There are several options for the rank method with `.rank()` including dense rank. Dense rank is where the rank immediately following those assigned to tied elements is given to the next highest element. For additional parameter options, refer to the Polars documentation: `https://pola-rs.github.io/polars/py-polars/html/reference/expressions/api/polars.Expr.rank.html`.

> **Tip**
>
> The `.over()` expression has to come after another aggregation or expression such as `.sum()` and `.rank()` in order to correctly execute operations on defined groups. As an example, this won't give you the expected result: `pl.col('column A').over('column B').sum()`.

There's more...

There are use cases for `join` and `explode` in the parameter for the `.over()` expression.

1. First, let's put the last result into a variable:

```
max_sales_rank = (
    df
    .group_by('Category', 'Subcategory')
    .agg(pl.col('Sales Amount').max().round().cast(pl.Int64).
alias('Max Sales Amt'))
    .with_columns(
        pl.col('Max Sales Amt').rank(descending=True).
over('Category').cast(pl.Int64).alias('Rank')
    )
    .filter(pl.col('Category').is_in(['Audio', 'Computers']))
    .sort(['Category', 'Rank'])
)
```

2. Now we'll get the list of three subcategories with the lowest max sales per category using `join`. One caveat is that using `join` can be memory intensive:

```
max_sales_rank.with_columns(
    pl.col('Subcategory')
    .sort_by('Max Sales Amt')
    .head(3)
    .over('Category', mapping_strategy='join')
    .alias('Lowest 3 Subcat per Cat')
)
```

The preceding code will return the following output:

shape: (9, 5)

Category	Subcategory	Max Sales Amt	Rank	Lowest 3 Subcat per Cat
str	str	i64	i64	list[str]
"Audio"	"Recording Pen"	2871	1	["MP4&MP3", "Bluetooth Headphones", "Recording Pen"]
"Audio"	"Bluetooth Headphones"	2250	2	["MP4&MP3", "Bluetooth Headphones", "Recording Pen"]
"Audio"	"MP4&MP3"	2095	3	["MP4&MP3", "Bluetooth Headphones", "Recording Pen"]
"Computers"	"Projectors & Screens"	19992	1	["Computers Accessories", "Printers, Scanners & Fax", "Monitors"]
"Computers"	"Laptops"	19485	2	["Computers Accessories", "Printers, Scanners & Fax", "Monitors"]
"Computers"	"Desktops"	14535	3	["Computers Accessories", "Printers, Scanners & Fax", "Monitors"]
"Computers"	"Monitors"	11425	4	["Computers Accessories", "Printers, Scanners & Fax", "Monitors"]
"Computers"	"Printers, Scanners & Fax"	2508	5	["Computers Accessories", "Printers, Scanners & Fax", "Monitors"]
"Computers"	"Computers Accessories"	2424	6	["Computers Accessories", "Printers, Scanners & Fax", "Monitors"]

Figure 4.33 – Rankings in a list column

3. Finally, you can use `explode` to transform the group of values into rows, whereas with `join`, we packed values into lists:

```
max_sales_rank.with_columns(
    pl.col('Subcategory')
    .sort_by('Max Sales Amt')
    .over('Category', mapping_strategy='explode')
    .alias('Subcategory Sorted by Max Sales Amt Ascending')
)
```

The preceding code will return the following output:

shape: (9, 5)

Category	Subcategory	Max Sales Amt	Rank	Subcategory Sorted by Max Sales Amt Ascending
str	str	i64	i64	str
"Audio"	"Recording Pen"	2871	1	"MP4&MP3"
"Audio"	"Bluetooth Headphones"	2250	2	"Bluetooth Headphones"
"Audio"	"MP4&MP3"	2095	3	"Recording Pen"
"Computers"	"Projectors & Screens"	19992	1	"Computers Accessories"
"Computers"	"Laptops"	19485	2	"Printers, Scanners & Fax"
"Computers"	"Desktops"	14535	3	"Monitors"
"Computers"	"Monitors"	11425	4	"Desktops"
"Computers"	"Printers, Scanners & Fax"	2508	5	"Laptops"
"Computers"	"Computers Accessories"	2424	6	"Projectors & Screens"

Figure 4.34 – Rankings in rows instead of in a list

You might wonder what the difference is between `group_to_rows` (the default option) and `explode`. Let's say I sorted the DataFrame differently before applying `.over()` with `explode`; we still get the same values in the new column, as seen in the preceding figure. It's simply flattening the values regardless of how they fit in the original DataFrame. You just need to be careful about how your DataFrame is sorted before using the `explode` option. The results of the new column no longer make sense:

```
(
    max_sales_rank
    .sort('Subcategory')
    .with_columns(
        pl.col('Subcategory')
        .sort_by('Max Sales Amt')
        .over('Category', mapping_strategy='explode')
        .alias('Subcategory Sorted by Max Sales Amt Ascending')
    )
)
```

The preceding code will return the following output. The output shows that the new column still gives you the same output as seen in the previous example.

shape: (9, 5)

Category	Subcategory	Max Sales Amt	Rank	Subcategory Sorted by Max Sales Amt Ascending
str	str	i64	i64	str
"Audio"	"Bluetooth Headphones"	2250	2	"MP4&MP3"
"Computers"	"Computers Accessories"	2424	6	"Bluetooth Headphones"
"Computers"	"Desktops"	14535	3	"Recording Pen"
"Computers"	"Laptops"	19485	2	"Computers Accessories"
"Audio"	"MP4&MP3"	2095	3	"Printers, Scanners & Fax"
"Computers"	"Monitors"	11425	4	"Monitors"
"Computers"	"Printers, Scanners & Fax"	2508	5	"Desktops"
"Computers"	"Projectors & Screens"	19992	1	"Laptops"
"Audio"	"Recording Pen"	2871	1	"Projectors & Screens"

Figure 4.35 – The same output despite the different sorting logic of the original DataFrame

See also

Please refer to these additional resources for learning more about window functions:

- `https://pola-rs.github.io/polars/py-polars/html/reference/expressions/api/polars.Expr.over.html#polars.Expr.over`

- `https://pola-rs.github.io/polars/user-guide/expressions/window/`

Applying UDFs

UDFs are functions defined by the user to encapsulate a block of code for reuse. Polars allows you to utilize UDFs to implement your logic in your code. The only caution is that once you use any of the methods explained in this recipe, you'll lose parallelization, and the operations are applied row by row. That potentially leads to slow performance, depending on the size of the dataset and the complexity of your code.

In this recipe, we'll cover how to utilize UDFs in Polars using the `.map_elements()` expression.

Getting ready

We'll use the Contoso dataset for this recipe as well. Run the following code to read the dataset:

```
df = pl.read_csv('../data/contoso_sales.csv', try_parse_dates=True)
```

How to do it...

Here are ways for how you apply UDFs:

1. Define a function that extracts the first name from a full name. Apply the function using `.map_elements()`:

```python
def get_first_name(full_name: str) -> str:
    return full_name.split(' ')[0]


df.select(
    'Customer Name',
    pl.col('Customer Name').map_elements(lambda el: get_first_name(el), return_dtype=pl.String).alias('Customer First Name')
).head()
```

The preceding code will return the following output:

shape: (5, 2)

Customer Name	Customer First Name
str	str
"Eric Kennedy"	"Eric"
"George Tooth"	"George"
"Caleb Greene"	"Caleb"
"Isaac Siddins"	"Isaac"
"Mike McQueen"	"Mike"

Figure 4.36 – The output of the map_elements expression

2. Use the same logic used in *step 1*, without defining a function beforehand. This results in the same output as the previous example:

```python
df.select(
    'Customer Name',
    pl.col('Customer Name').map_elements(lambda el: el.split(' ')[0], return_dtype=pl.String).alias('Customer First Name')
).head()
```

3. Apply an `if else` statement using `.map_elements()`:

```python
def age_to_range(age: int) -> str:
    if age < 18:
        return '~17'
    elif age <= 30:
        return '18~30'
```

```
    elif age <= 50:
        return '31~50'
    elif age <= 70:
        return '50~70'
    else:
        return '71~'

df.select(
    'Customer Age',
    pl.col('Customer Age').map_elements(lambda el: age_to_
range(el), return_dtype=pl.String).alias('Age Range')
).head()
```

The preceding code will return the following output:

shape: (5, 2)

Customer Age	Age Range
i64	str
47	"31~50"
30	"18~30"
59	"50~70"
25	"18~30"
56	"50~70"

Figure 4.37 – An if else statement using the map_elements expression

How it works...

If you're familiar with pandas, `.map_elements()` is the equivalent of `.apply()` in pandas. It allows you to define and apply your custom function. There are a few other variations of `.map_elements()` in Polars, such as `.map_batches()`, `.map_groups()`, and `.map_rows()`, but I won't go over the details of each. For more information about how they work, refer to the Polars' documentation: `https://pola-rs.github.io/polars/user-guide/expressions/user-defined-functions/`.

> **Important**
>
> As explained in the introduction, although UDFs provide convenience and usefulness in implementing your logic, they kill parallelization. So, the performance of your code suffers a lot depending on the volume of the dataset and the complexity of your logic. The rule of thumb is *not* to use UDFs and, instead, implement with expressions as much as possible. Given the flexibility that Polars gives you, it's usually the case that you don't need to use UDFs in the first place and enjoy the blazing-fast performance with Polars' expressions instead.

There's more...

The examples just demonstrated can be implemented using Polars' existing methods and expressions.

Let's extract the first name:

```
df.select(
    'Customer Name',
    pl.col('Customer Name').str.split(' ').list.first().
alias('Customer First Name')
)
```

Note that `.str` is used to access expressions for string manipulations and `.list` is used for accessing list operations. We'll cover both in *Chapter 6, Performing String Manipulations*, and *Chapter 7, Working with List and Array Operations*.

Your code will look like the following, replacing the `if else` logic with Polars' built-in expressions:

```
df.select(
    'Customer Age',
    pl.when(pl.col('Customer Age')<18).then(pl.lit('~17'))
    .when(pl.col('Customer Age')<=30).then(pl.lit('18~30'))
    .when(pl.col('Customer Age')<=50).then(pl.lit('31~50'))
    .when(pl.col('Customer Age')<=70).then(pl.lit('51~70'))
    .when(pl.col('Customer Age')>70).then(pl.lit('71~'))
    .alias('Age Range')
)
```

You can even benchmark the difference in performance using a magic command in Jupyter Notebook: `%%timeit`. You just add it at the beginning of your code in the cell. Here's an example showing that Polars' expressions are faster, even with our example that's simple in logic:

```
%%timeit
df.select(
    'Customer Name',
    pl.col('Customer Name').map_elements(lambda el: el.split(' ')[0],
return_dtype=pl.String).alias('Customer First Name')
).head()
```

```
>> 3.75 ms ± 116 µs per loop (mean ± std. dev. of 7 runs, 100 loops
each)
```

```
%%timeit
df.select(
    'Customer Name',
```

```
    pl.col('Customer Name').str.split(' ').list.first().
alias('Customer First Name')
).head()
```

```
>> 943 µs ± 20.5 µs per loop (mean ± std. dev. of 7 runs, 1,000 loops
each)
```

See also

To learn more about using UDFs in Polars, please refer to the following resources:

- https://pola-rs.github.io/polars/user-guide/expressions/user-defined-functions/

- https://pola-rs.github.io/polars/py-polars/html/reference/expressions/api/polars.Expr.map_elements.html

- https://pola-rs.github.io/polars/py-polars/html/reference/expressions/api/polars.Expr.map_batches.html

- https://docs.pola.rs/api/python/stable/reference/expressions/api/polars.map_groups.html

Using SQL for data transformations

SQL is an essential tool in data analysis and transformations. Many data pipelines you see at workplaces are written in SQL and most data professionals are accustomed to using SQL for analytics work.

The good news is that you can use SQL in Python Polars as well. This opens the door for those who might not be as familiar with a DataFrame library. In this recipe, we'll cover how to configure Polars to use SQL and how you can implement simple SQL queries such as aggregations.

Getting ready

We'll use the Contoso dataset for this recipe as well. Run the following code to read the dataset:

```
df = pl.read_csv('../data/contoso_sales.csv', try_parse_dates=True)
```

How to do it...

Here's how to use SQL in Polars:

1. Define the SQL context and register your DataFrame:

    ```
    ctx = pl.SQLContext(eager=True)
    ctx.register('df', df)
    ```

2. Create a simple query and execute it:

```
ctx.execute(
    """
    select
      `Customer Name`,
      Brand,
      Category
    from df limit 5
    """
)
```

The preceding code will return the following output:

shape: (5, 3)

Customer Name	Brand	Category
str	str	str
"Eric Kennedy"	"Contoso"	"Audio"
"George Tooth"	"Contoso"	"Audio"
"Caleb Greene"	"Contoso"	"Audio"
"Isaac Siddins"	"Contoso"	"Audio"
"Mike McQueen"	"Contoso"	"Audio"

Figure 4.38 – The result of a SQL query

3. Apply a group by aggregation, showing the top five average quantities by brand:

```
ctx.execute(
    """
    select
      Brand,
      avg(Quantity) as `Avg Quantity`
    from df
    group by
      Brand
    order by
      `Avg Quantity` desc
    limit 5
    """
)
```

The preceding code will return the following output:

shape: (5, 2)

Brand	Avg Quantity
str	f64
"Fabrikam"	3.225532
"Northwind Trad...	3.222222
"Wide World Imp...	3.193811
"Fabrikam "	3.192308
"Southridge Vid...	3.189509

Figure 4.39 – Top five average quantities by brand

4. Run a SQL query in one go, showing the first five rows for a few selected columns using a LazyFrame:

```
pl.SQLContext(lf=df.lazy()).execute(
    """
        select
            Brand,
            Category
        from lf
        limit 5
    """
).collect()
```

The preceding code will return the following output:

shape: (5, 2)

Brand	Category
str	str
"Contoso"	"Audio"
"Contoso"	"Audio"
"Contoso"	"Audio"
"Contoso"	"Audio"
"Contoso"	"Audio"

Figure 4.40 – The first five rows with the Brand and Category columns selected

How it works...

By default, your SQL query returns a LazyFrame as its output. You can specify to return a DataFrame by specifying a parameter either when you define your SQL context or execute your query.

Also, know that when your column name contains a space, you need to use a special character, ` (a backtick or grave accent).

One thing to keep in mind when using SQL in Polars is that, as of this writing, not all data operations are available such as rank and row number. I suggest you read Polars' documentation to learn more about what Polars SQL can and cannot do: `https://pola-rs.github.io/polars/user-guide/sql/intro/`.

> **Tip**
> You can use a library such as DuckDB to run SQL queries on top of Polars' DataFrames. It basically overcomes the limitation of available SQL transformations in Polars. We'll cover how to use other Python libraries in combination with Polars in *Chapter 10, Interoperability with Other Python Libraries.*

See also

Please refer to these resources to learn more about using SQL in Polars:

- `https://docs.pola.rs/user-guide/sql/intro/`
- `https://pola-rs.github.io/polars/user-guide/sql/select/`

5

Handling Missing Data

In data analysis, data science, and data engineering, the majority of time is spent doing data manipulations and cleaning. Your data could be very messy in that it contains a lot of missing data that you need to treat with care. To compute whatever you need, you may need to identify missing data and decide what to do with it.

There are two approaches to handling missing data. One is to substitute missing data with alternate values. Another way is to simply drop records that contain missing data. However, your decision to handle missing data should align with your end goal. That helps identify the appropriate approach as well as values with which you may want to replace missing data.

We'll cover `null` and **Not a Number (NaN)** values in this chapter. Polars treats them differently and NaN values are technically a type of floating point data rather than missing data. That also means that there are different methods and expressions for `null` and NaN.

You'll learn ways to handle missing data, from identifying missing records to filling them with appropriate values. You'll be equipped with the knowledge to deal with situations that require you to be creative in handling missing data.

In this chapter, we're going to cover the following main recipes:

- Identifying missing data
- Deleting rows and columns containing missing data
- Filling in missing data

Technical requirements

You can download the datasets and code from the GitHub repository at the following links:

- Data: `https://github.com/PacktPublishing/Polars-Cookbook/tree/main/data`
- Code: `https://github.com/PacktPublishing/Polars-Cookbook/tree/main/Chapter05`

It is assumed that you have installed the Polars library in your Python environment:

```
>>> pip install polars
```

It is also assumed that you have imported it into your code:

```
import polars as pl
```

Identifying missing data

The first step in handling missing data is to identify whether there is missing data and how many instances of it you have in your data. Polars provides several ways to accomplish that.

Getting ready

We'll be using the NumPy library to generate NaN values. Note that you can still generate NaN values in native Python with code such as `float('nan')`.

Install numpy with the following command if you haven't already as Polars' dependency:

```
>>> pip install numpy
```

We'll be using a dataset that we manually create. Make sure to run the following code before proceeding to the next steps:

```
from datetime import date
import numpy as np

date_col = pl.date_range(date(2023, 1, 1), date(2023, 1, 15), '1d',
eager=True)
avg_temp_c_list = [-3,None,6,-1,np.nan,6,4,None,1,2,np.nan,7,9,-
2,None]
df = pl.DataFrame({
    'date': date_col,
    'avg_temp_celsius': avg_temp_c_list
}, strict=False)
```

Alternatively, you can read a CSV file I created from the preceding code:

```
df = pl.read_csv('../data/temperatures.csv')
```

How to do it...

Here are the ways to identify missing data. We will cover how to do this for `null` values first and then NaN:

1. Check how many `null` values exist in each column in a DataFrame:

    ```
    df.null_count()
    ```

    The preceding code will return the following output:

    shape: (1, 2)

date	avg_temp_celsius
u32	u32
0	3

    Figure 5.1 – Null counts in each column

2. Check `null` values in a selected column:

    ```
    df.select('avg_temp_celsius').null_count()
    ```

    The preceding code will return in the following output:

    shape: (1, 1)

avg_temp_celsius
u32
3

    Figure 5.2 – Null counts in a particular column

3. Check `null` values in multiple columns:

    ```
    df.select('date', 'avg_temp_celsius').null_count()
    ```

    The preceding code will return the following output:

    shape: (1, 2)

date	avg_temp_celsius
u32	u32
0	3

    Figure 5.3 – Null counts in multiple columns

4. Use an expression to check null values:

    One way is to use .null_count() as an expression:

    ```
    df.select(pl.col('avg_temp_celsius').null_count())
    ```

    Another way is to use the combination of .is_null() and .sum():

    ```
    df.select(
        pl.col('avg_temp_celsius')
        .is_null()
        .sum()
    )
    ```

    The two preceding blocks of code will return the following output:

    shape: (1, 1)

avg_temp_celsius
u32
3

    Figure 5.4 – Null counts in a particular column with expressions

    You could also utilize the .filter() method and count the rows that are null at the end:

    ```
    (
        df
        .filter(pl.col('avg_temp_celsius').is_null())
        .select(pl.len())
    )
    ```

    The preceding code will return the following output:

    shape: (1, 1)

len
u32
3

    Figure 5.5 – Null counts in a column using the filter expression

    Another example of counting null values with .filter() is the following:

    ```
    df.filter(pl.col('avg_temp_celsius').is_null()).shape[0]
    ```

    Here's the output:

    ```
    >> 3
    ```

5. Check NaN values in a DataFrame for selected columns:

```
df.select(
    pl.col('avg_temp_celsius')
    .is_nan()
    .sum()
)
```

The preceding code will return the following output:

shape: (1, 1)

avg_temp_celsius
u32
2

Figure 5.6 – NaN counts using .is_nan() and .sum()

Here's the code to filter out NaN values using the `.is_nan()` method:

```
(
    df
    .filter(pl.col('avg_temp_celsius').is_nan())
    .select(pl.len())
)
```

The preceding code will return the following output:

shape: (1, 1)

len
u32
2

Figure 5.7 – NaN counts using .filter()

How it works...

As you have noticed, methods or expressions for `null` values won't affect NaN values, and vice versa. You must use different methods and expressions to identify `null` or NaN values. For instance, there is no method such as `.null_count()` for NaN. You can utilize `.is_nan()` in combination with other methods or expressions.

You might wonder how the combination of `.is_null()` or `.is_nan()` and `.sum()` returns the total null/NaN count. It's because `.is_null()` and `.is_nan()` return Boolean values. `True` gets interpreted as the value of `1`, whereas `False` returns `0`. That means you'd get the total count if you sum it all up.

> **Tip**
>
> You can utilize expressions such as `.is_not_null()` and `.is_not_nan()` to get non-null/non-NaN values.

> **Note**
>
> Starting from Polars version 1.0.0, the behavior of the Series constructor has been updated to be stricter by default. The reason behind this change is that strict construction is more efficient than non-strict construction. Therefore, if you want to pass values of different data types when creating a Series, you'll need to set the `strict` parameter to *False*. This is why, when we created the DataFrame for this recipe, we set the `strict` parameter to *False*.

See also

Please refer to these additional resources to learn more about identifying missing data:

- `https://docs.pola.rs/user-guide/expressions/missing-data/`
- `https://stuffbyyuki.com/handling-missing-values-in-polars/`

Deleting rows and columns containing missing data

One of the ways to manage missing data is to simply drop records that contain missing data. You could also drop columns if you decide they're not useful given how many rows are missing.

In this recipe, we'll cover how to delete rows and columns that contain missing data.

Getting ready

Read the same dataset that we used in the previous recipe:

```
df = pl.read_csv('temperatures.csv')
```

How to do it...

Here are the ways to delete rows and columns that contain missing data:

1. Delete rows that contain `null` values in a whole DataFrame using `.drop_nulls()`. Apply the `.null_count()` method to check that it worked:

```
df.drop_nulls().null_count()
```

Here's the output of the preceding code.

shape: (1, 2)

date	avg_temp_celsius
u32	u32
0	0

Figure 5.8 – DataFrame after dropping nulls

2. Delete rows with `null` values for selected columns:

```
df.select(
    pl.col('avg_temp_celsius')
    .drop_nulls()
    .null_count()
)
```

The preceding code will return the following output:

shape: (1, 1)

avg_temp_celsius
u32
0

Figure 5.9 – A column after dropping null values

You can do the same thing using `.is_not_null()`. The following code will return a DataFrame without `null` values in the `avg_temp_celsius` column:

```
df.filter(pl.col('avg_temp_celsius').is_not_null())
```

3. Drop columns if they contain at least a `null` value. We first identify columns that contain a `null` value and then drop them with `.drop()`:

```
cols_to_drop = [column for column in df.columns if df.select(pl.
col(column).is_null().any())[0,0]]
df.drop(cols_to_drop).columns
```

The preceding code will return the following output:

```
>> ['date']
```

4. You can also accomplish the same thing done in *step 3* with the `.item()` method. This approach is more descriptive and easier to read:

```
cols_to_drop = [column for column in df.columns if df.select(pl.
col(column).is_null().any()).item()]
df.drop(cols_to_drop).columns
```

The preceding code will return the following output:

```
>> ['date']
```

5. Drop rows that contain NaN values using `.drop_nans()` and then check the output:

```
df.select(
    pl.col('avg_temp_celsius')
    .drop_nans()
    .is_nan()
    .sum()
)
```

The preceding code will return the following output:

shape: (1, 1)

avg_temp_celsius
u32
0

Figure 5.10 – No NaN values in the column

You can do the same with `.is_not_nan()`. This code will return the dropped DataFrame NaN values:

```
df.filter(pl.col('avg_temp_celsius').is_not_nan())
```

6. Drop columns that contain NaN values. We'll create a similar logic as we built in *step 3*:

```
import polars.selectors as cs
cols_to_drop = df.select(cs.float().is_nan().any()).columns
df.drop(cols_to_drop).columns
```

The preceding code will return the following output:

```
>> ['date']
```

You can also retain necessary columns instead of dropping the columns we don't need:

```
df.select(pl.exclude(cols_to_drop)).columns
```

The preceding code will return the following output:

```
>> ['date']
```

How it works...

There are two ways to drop unnecessary rows and columns. One is to delete what's not needed. The other is to keep what's necessary. Depending on your logic and requirements, one way may make more sense. The good thing is that Polars provides you with both ways and you have the flexibility in how you build your data processing pipelines.

In a few of the preceding examples, I used [0,0], with which I specified the index of the row and the column. Like pandas, Polars provides a way to choose rows and columns using indexes. Inside the brackets, the number on the left is for the row, and the one on the right is for the column. The .item() method works in a similar way. You can specify the row and column index as its parameters. If your DataFrame has the shape (1,1), then using .item() without specifying any parameter works exactly the same as using [0, 0].

The .any() expression is to check whether any of the values in the column is true. You could use the .all() expression if you want to check whether all the values in the column are true.

> **Note**
> One thing worth noting is that when you check NaN values with the .is_nan() expression, you can only check columns of the float data type. Otherwise, it'll throw an error. As mentioned in the introduction of this chapter, NaN values are not considered missing data, but they are values of the float data type.

There's more...

If you want to treat null and NaN values in the same way simultaneously, you can first convert NaN values to null values and then remove them altogether. This would be useful when NaN values can't be dealt with. We'll learn more about how to fill in missing data in the next recipe.

```
df.fill_nan(None).drop_nulls()
```

The preceding code will return the following output:

shape: (10, 2)

date	avg_temp_celsius
str	f64
"2023-01-01"	-3.0
"2023-01-03"	6.0
"2023-01-04"	-1.0
"2023-01-06"	6.0
"2023-01-07"	4.0
"2023-01-09"	1.0
"2023-01-10"	2.0
"2023-01-12"	7.0
"2023-01-13"	9.0
"2023-01-14"	-2.0

Figure 5.11 – No NaN values in the column

See also

These resources are helpful for learning more about deleting missing data:

- `https://stuffbyyuki.com/handling-missing-values-in-polars/`
- `https://docs.pola.rs/user-guide/expressions/missing-data/`
- `https://pola-rs.github.io/polars/py-polars/html/reference/dataframe/api/polars.DataFrame.drop.html`
- `https://pola-rs.github.io/polars/py-polars/html/reference/expressions/api/polars.Expr.any.html#polars.Expr.any`
- `https://pola-rs.github.io/polars/py-polars/html/reference/expressions/api/polars.exclude.html#polars-exclude`

Filling in missing data

One way for you to handle missing data is by filling it with substitutions. This is also called **imputation**. If you're building a machine learning model or conducting a statistical test, how you fill your missing data can affect your model output. Knowing the various ways of filling in missing data gives you the options from which you can choose the best approach for your particular use case.

In this recipe, we'll look at how to fill missing data with a constant value, strategy, interpolation, and expressions.

Getting ready

We'll be using the same temperature dataset we've used throughout this chapter. Run the following code to read the CSV file:

```
df = pl.read_csv('../data/temperatures.csv')
```

> **Note**
>
> We will only cover how to fill `null` values and won't cover how to fill NaN values as the functionality of the methods and expressions are also available for NaN values. For instance, the `.fill_null()` expression is available as `.fill_nan()` for NaN values.

How to do it...

Here are the ways to fill in missing data:

1. Fill in missing data with a constant value:

    ```
    df.select(
        pl.col('avg_temp_celsius'),
        avg_temp_nulls_filled=pl.col('avg_temp_celsius').fill_
    null(pl.lit('1'))
    )
    ```

The preceding code will produce the following output:

shape: (15, 2)

avg_temp_celsius	avg_temp_nulls_filled
f64	str
-3.0	"-3.0"
null	"1"
6.0	"6.0"
-1.0	"-1.0"
NaN	"NaN"
6.0	"6.0"
4.0	"4.0"
null	"1"
1.0	"1.0"
2.0	"2.0"
NaN	"NaN"
7.0	"7.0"
9.0	"9.0"
-2.0	"-2.0"
null	"1"

Figure 5.12 – Null values are replaced with 1

2. Fill in missing data using a strategy:

```
df.select(
    pl.col('avg_temp_celsius'),
    forward_filled=pl.col('avg_temp_celsius').fill_
null(strategy='forward'),
    backward_filled=pl.col('avg_temp_celsius').fill_
null(strategy='backward'),
    mean_filled=pl.col('avg_temp_celsius').fill_
null(strategy='mean'),
    min_filled=pl.col('avg_temp_celsius').fill_
null(strategy='min'),
    max_filled=pl.col('avg_temp_celsius').fill_
null(strategy='max'),
)
```

The preceding code will return the following output:

shape: (15, 6)

avg_temp_celsius	forward_filled	backward_filled	mean_filled	min_filled	max_filled
f64	f64	f64	f64	f64	f64
-3.0	-3.0	-3.0	-3.0	-3.0	-3.0
null	-3.0	6.0	NaN	-3.0	9.0
6.0	6.0	6.0	6.0	6.0	6.0
-1.0	-1.0	-1.0	-1.0	-1.0	-1.0
NaN	NaN	NaN	NaN	NaN	NaN
6.0	6.0	6.0	6.0	6.0	6.0
4.0	4.0	4.0	4.0	4.0	4.0
null	4.0	1.0	NaN	-3.0	9.0
1.0	1.0	1.0	1.0	1.0	1.0
2.0	2.0	2.0	2.0	2.0	2.0
NaN	NaN	NaN	NaN	NaN	NaN
7.0	7.0	7.0	7.0	7.0	7.0
9.0	9.0	9.0	9.0	9.0	9.0
-2.0	-2.0	-2.0	-2.0	-2.0	-2.0
null	-2.0	null	NaN	-3.0	9.0

Figure 5.13 – Null values are replaced with strategies

3. Fill in missing data using interpolation:

```
df.select(
    pl.col('avg_temp_celsius'),
    interpolated_linear=pl.col('avg_temp_celsius').
interpolate(),
    interpolated_nearest=pl.col('avg_temp_celsius').
interpolate(method='nearest')
)
```

The preceding code will produce the following output:

shape: (15, 3)

avg_temp_celsius	interpolated_linear	interpolated_nearest
f64	f64	f64
-3.0	-3.0	-3.0
null	1.5	6.0
6.0	6.0	6.0
-1.0	-1.0	-1.0
NaN	NaN	NaN
6.0	6.0	6.0
4.0	4.0	4.0
null	2.5	1.0
1.0	1.0	1.0
2.0	2.0	2.0
NaN	NaN	NaN
7.0	7.0	7.0
9.0	9.0	9.0
-2.0	-2.0	-2.0
null	null	null

Figure 5.14 – Null values are replaced with interpolation

4. Fill in missing data using expressions:

```
df.select(
    'avg_temp_celsius',
    avg_temp_median=pl.col('avg_temp_celsius')
        .fill_null(
            pl.col('avg_temp_celsius').median()
        ),
    avg_temp_max_minus_min=pl.col('avg_temp_celsius')
        .fill_null(
            pl.col('avg_temp_celsius').max() - pl.col('avg_temp_
celsius').min()
        )
)
```

The preceding code will return the following output:

shape: (15, 3)

avg_temp_celsius	avg_temp_median	avg_temp_max_minus_min
f64	f64	f64
-3.0	-3.0	-3.0
null	5.0	12.0
6.0	6.0	6.0
-1.0	-1.0	-1.0
NaN	NaN	NaN
6.0	6.0	6.0
4.0	4.0	4.0
null	5.0	12.0
1.0	1.0	1.0
2.0	2.0	2.0
NaN	NaN	NaN
7.0	7.0	7.0
9.0	9.0	9.0
-2.0	-2.0	-2.0
null	5.0	12.0

Figure 5.15 – Null values are replaced with expressions

How it works...

`pl.lit()` is what you use to define a constant value or a literal. Otherwise, Polars might recognize it as a column name. It's better to be explicit in what you write.

We demonstrated filling missing data with a strategy with calculations such as `mean`, `min`, `max`, `forward`, and `backward`. You can also fill missing data with zeros and ones with `.fill_null()`. If you're interested in learning more about what other strategies are available, please refer to the *See also* section.

Interpolation is an estimation technique in mathematics that creates new data points based on the range of existing data points. You can access it using the `.interpolate()` expression. This only works on `null` values and won't work on NaN values.

Expressions can be used to define how you impute your missing data. You can imagine how this opens doors for more advanced ways to manage your missing data.

> **Note**
>
> Another option for filling in missing data is to use machine learning algorithms such as k-nearest neighbors and XGBoost. Data imputation with machine learning is out of the scope of this book; however, it's an effective way to impute data because machine learning models are able to capture patterns that we humans wouldn't be able to. I recommend looking at another Python library such as scikit-learn to learn more: `https://scikit-learn.org/`.

There's more...

You can use the `.forward_fill()` and `.backward_fill()` expressions instead of strategies in `.fill_null()`.

First, let's create a DataFrame for a demonstration:

```
df = pl.DataFrame(
    {'values': [1,2,None,None,None,3,4,None,5]}
)
```

Here is an example of using `.forward_fill()` and `.backward_fill()`:

```
df.select(
    'values',
    forward_fill=pl.col('values').forward_fill(),
    forward_fill_1=pl.col('values').forward_fill(limit=1),
    backward_fill=pl.col('values').backward_fill(),
    backward_fill_2=pl.col('values').backward_fill(limit=2),
)
```

The preceding code will return the following output:

shape: (9, 5)

values	forward_fill	forward_fill_1	backward_fill	backward_fill_2
i64	i64	i64	i64	i64
1	1	1	1	1
2	2	2	2	2
null	2	2	3	null
null	2	null	3	3
null	2	null	3	3
3	3	3	3	3
4	4	4	4	4
null	4	4	5	5
5	5	5	5	5

Figure 5.16 – Null values are replaced with forward_fill and backward_fill expressions

> **Tip**
>
> If you want to use methods or expressions for NaN values but they are only available for `null` values, you can always convert NaN values to `null` values using `.fill_nan(None)`.

See also

You can refer to the following resources to learn more about filling in missing data:

- `https://stuffbyyuki.com/handling-missing-values-in-polars/`
- `https://pola-rs.github.io/polars/py-polars/html/reference/expressions/api/polars.Expr.fill_null.html`
- `https://pola-rs.github.io/polars/py-polars/html/reference/expressions/api/polars.Expr.fill_nan.html`
- `https://pola-rs.github.io/polars/py-polars/html/reference/expressions/api/polars.Expr.interpolate.html`
- `https://scikit-learn.org/stable/modules/generated/sklearn.impute.KNNImputer.html`
- `https://pola-rs.github.io/polars/py-polars/html/reference/expressions/api/polars.Expr.forward_fill.html`
- `https://pola-rs.github.io/polars/py-polars/html/reference/expressions/api/polars.Expr.backward_fill.html`

6

Performing String Manipulations

String manipulations are an important technique in data analysis and wrangling workflows. Your data may not be in the most organized format, or it may need extra tweaks to conduct further analysis. String manipulation techniques essentially allow us to extract meaningful information from strings or text data. There are several techniques available in Python Polars for performing string manipulations.

This chapter includes the following recipes:

- Filtering strings
- Converting strings into date, datetime, or time
- Extracting substrings
- Cleaning strings
- Splitting strings into lists and structs
- Concatenating and combining strings

Technical requirements

You can download the datasets and code from the GitHub repository:

- Data: `https://github.com/PacktPublishing/Polars-Cookbook/tree/main/data`
- Code: `https://github.com/PacktPublishing/Polars-Cookbook/tree/main/Chapter06`

It is assumed that you have installed the Polars library in your Python environment:

```
>>> pip install polars
```

It is also assumed that you have imported it into your code:

```
import polars as pl
```

We'll be using the Google Play Store Reviews dataset throughout this chapter, which can be found here: https://github.com/PacktPublishing/Polars-Cookbook/blob/main/data/google_store_reviews.csv.

Read the dataset into a DataFrame:

```
df = pl.read_csv('../data/google_store_reviews.csv')
```

Let's change the length of a string that Polars will show in a DataFrame. For Jupyter notebooks, you need to change the environmental variable directly:

```
import os
os.environ['POLARS_FMT_STR_LEN'] = str(50)
```

When you're using Python scripts, you just need to declare the following code:

```
pl.Config.set_fmt_str_lengths(50)
```

Always check your dataset before conducting any analysis or transformations to get an idea of what the data looks like:

```
df.head()
```

Filtering strings

One of the most common tasks when working with strings is to filter strings. By capturing patterns found in strings, you can filter records in your DataFrame. This allows you to apply business logic on strings to keep only what you need or discard what you don't need for your analysis.

In this recipe, we'll cover how to filter strings using string methods such as .str.starts_with(), .str.ends_with(), and .str.contains().

How to do it...

Here are five ways in which you can filter strings:

- Filter a string based on the characters that it starts with. This is an example of a case-sensitive substring match:

```
(
    df
    .filter(pl.col('content').str.starts_with('Very'))
```

```
        .select('content')
        .head()
)
```

The preceding code will return the following output:

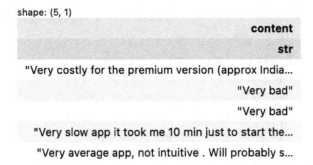

Figure 6.1 – The first five rows where the content starts with Very

- Filter a string based on the characters it ends with:

```
(
    df
    .filter(pl.col('userName').str.ends_with('Smith'))
    .select('userName')
    .head()
)
```

The preceding code will return the following output:

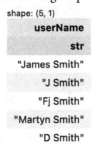

Figure 6.2 – The first five rows where the usernames end with Smith

- Filter a string based on the characters it contains:

```
(
    df
    .filter(pl.col('content').str.contains('happy',
literal=True))
```

```
        .select('content')
        .head()
)
```

The preceding code will return the following output:

shape: (5, 1)

content
str
"I love this app, but I do have one major gripe – ...
"Not happy, app just asked me to 'sign in' and now...
"Will be happy if this app comes with time duratio...
"V usefull app i love it v much I use it daily wor...
"I was super happy to download this app but that I...

Figure 6.3 – The first five rows where the content contains the word happy

- Filter a string based on the character it contains. This time, the condition is specified with **regular expressions (regex)**:

```
(
    df
    .filter(pl.col('content').str.contains(r'very happy|best
app|I love'))
    .select('content')
    .head()
)
```

The preceding code will return the following output:

shape: (5, 1)

content
str
"I love this app, but I do have one major gripe – ...
"Why are random items popping up on our Grocery Li...
"I love using this app however when I installed it...
"There are certain things I love like the fact tha...
"I love this app, but recently the app keeps crash...

Figure 6.4 – The first five rows where the content contains words specified by the regex

- Use `.str.contains_any()` to check if a string contains any of the values you specify:

```
(
    df
    .filter(pl.col('content').str.contains_any(['happy', 'love',
'best']))
    .select('content')
    .height
)
```

The output of the preceding code is as follows:

```
>> 1237
```

How it works...

There are many string methods available under the `str` namespace in Polars. Out of those string methods, the one we just covered is useful in filtering strings. When it comes to using regex, it's a powerful tool with which you can apply complex logic for filtering and matching strings as per business requirements.

The `.str.contains()` method has the parameter to choose either literal values or regex. It defaults to `False`, meaning it'd use regex by default unless you explicitly specify `literal=True`. It is recommended that you set it to `True` if you don't need regex, since regex is more computationally intensive than simple string searches. The strict parameter is another useful thing to keep in mind. Polars will raise exceptions if the regex pattern is not valid. It'll mask out with a null value otherwise.

There's more...

Here are a few other examples of filtering strings using other string methods.

To filter based on the count of records that match the given logic, you would do the following:

```
(
    df
    .filter(pl.col('content').str.count_matches(r'very happy|best
app|I love') > 2)
    .select('content')
)
```

The preceding code will return the following output:

shape: (4, 1)

content
str
"I love it :D I met great and fun people and this ...
"I love Habitica! I've used it for several years, ...
"A lot of work was put into this. I love the idea ...
"Very nice app I downloaded many app but it is the...

Figure 6.5 – Records that match the given logic

To show only those whose names are longer than 10 characters, use the following:

```
(
    df
    .filter(pl.col('userName').str.len_chars() > 10)
    .select('userName')
    .head()
)
```

The preceding code will return the following output:

shape: (5, 1)

userName
str
"Sudhakar .S"
"SKGflorida@bellsouth.net DAVID S"
"Louann Stoker"
"Jon Clemens"
"I Dewa Gede Nopi Ariana"

Figure 6.6 – Names longer than 10 characters

The .len_chars() method was used in the preceding example. However, if you'd like to get each string in the number of bytes, you can use the .len_bytes() method instead. Note that the lengths in bytes and characters are not the same.

See also

To learn more about string methods for filtering, you can refer to these additional resources:

- `https://pola-rs.github.io/polars/py-polars/html/reference/expressions/api/polars.Expr.str.starts_with.html`

- `https://docs.pola.rs/api/python/stable/reference/expressions/api/polars.Expr.str.ends_with.html`

- `https://docs.pola.rs/api/python/stable/reference/expressions/api/polars.Expr.str.contains.html`

- `https://docs.rs/regex/latest/regex/#grouping-and-flags`

Converting strings into date, time, and datetime

When you read in a dataset, you may find that columns that are supposed to be of the date, time, or datetime data type are read as the string data type. You can try to fix that at the source or within your method such as `.read_csv()`, but there are cases in which you don't have that flexibility. The good news is that Polars has built-in methods to help convert strings into date, time, and datetime values. You can apply those methods after reading in a dataset.

In this recipe, we'll look at how we can utilize string methods such as `.str.to_date()`, `.str.to_time()`, `.str.to_datetime()`, and `.str.strptime()`.

How to do it...

Here's how to utilize the methods to convert strings to date, time, or datetime values:

1. Convert a string column to a date column:

```
df.select(
    'at',
    pl.col('at').str.to_date(format='%Y-%m-%d %H:%M:%S').
alias('at(date)')
).head()
```

The preceding code will return the following output:

shape: (5, 2)

at	at(date)
str	date
"2020-10-27 21:24:41"	2020-10-27
"2020-10-27 14:03:28"	2020-10-27
"2020-10-27 08:18:40"	2020-10-27
"2020-10-26 13:28:07"	2020-10-26
"2020-10-26 06:10:50"	2020-10-26

Figure 6.7 – A string column converted to a date column

2. Convert a string column to a time column:

```
df.select(
    'at',
    pl.col('at').str.to_time(format='%Y-%m-%d %H:%M:%S').
alias('at(time)')
).head()
```

The preceding code will return the following output:

shape: (5, 2)

at	at(time)
str	time
"2020-10-27 21:24:41"	21:24:41
"2020-10-27 14:03:28"	14:03:28
"2020-10-27 08:18:40"	08:18:40
"2020-10-26 13:28:07"	13:28:07
"2020-10-26 06:10:50"	06:10:50

Figure 6.8 – A string column converted to a time column

3. Convert a string column to a datetime column:

```
df.select(
    'at',
    pl.col('at').str.to_datetime(format='%Y-%m-%d %H:%M:%S').
alias('at(datetime)')
).head()
```

The preceding code will return the following output:

shape: (5, 2)

at	at(datetime)
str	datetime[µs]
"2020-10-27 21:24:41"	2020-10-27 21:24:41
"2020-10-27 14:03:28"	2020-10-27 14:03:28
"2020-10-27 08:18:40"	2020-10-27 08:18:40
"2020-10-26 13:28:07"	2020-10-26 13:28:07
"2020-10-26 06:10:50"	2020-10-26 06:10:50

Figure 6.9 – A string column converted to a datetime column

4. Convert a string to date, time, or datetime with `.str.strptime()`:

```
df.select(
    'at',
    pl.col('at').str.strptime(pl.Date, '%Y-%m-%d %H:%M:%S').
alias('at(date)'),
    pl.col('at').str.strptime(pl.Time, '%Y-%m-%d %H:%M:%S').
alias('at(time)'),
    pl.col('at').str.strptime(pl.Datetime, '%Y-%m-%d %H:%M:%S').
alias('at(datetime)')
).head()
```

The preceding code will return the following output:

shape: (5, 4)

at	at(date)	at(time)	at(datetime)
str	date	time	datetime[µs]
"2020-10-27 21:24:41"	2020-10-27	21:24:41	2020-10-27 21:24:41
"2020-10-27 14:03:28"	2020-10-27	14:03:28	2020-10-27 14:03:28
"2020-10-27 08:18:40"	2020-10-27	08:18:40	2020-10-27 08:18:40
"2020-10-26 13:28:07"	2020-10-26	13:28:07	2020-10-26 13:28:07
"2020-10-26 06:10:50"	2020-10-26	06:10:50	2020-10-26 06:10:50

Figure 6.10 – A string column converted to a date, time, or datetime column

How it works...

All of the methods introduced in the preceding section need the date format to be specified. You can read more about the formatting syntax in this Rust documentation: `https://docs.rs/chrono/latest/chrono/format/strftime/index.html`. In our dataset, the at column includes year, month, day, hour, minute, and second. The format we needed to specify was `%Y-%m-%d %H:%M:%S`.

The .str.strptime() method is the all-rounder that can convert a string to any date type format, unlike the other methods we covered in the recipe. However, you need to specify the target data type within the method in addition to the input date formatting. With that said, using more descriptive, built-in methods such as .str.to_date(), .str.to_time(), and .str.to_datetime() is likely to be more appropriate unless you have a good reason to use .str.strptime().

See also

These are additional resources for you to explore more in detail:

- https://pola-rs.github.io/polars/py-polars/html/reference/ expressions/api/polars.Expr.str.to_date.html

- https://pola-rs.github.io/polars/py-polars/html/reference/ expressions/api/polars.Expr.str.to_time.html

- https://pola-rs.github.io/polars/py-polars/html/reference/ expressions/api/polars.Expr.str.to_datetime.html

- https://pola-rs.github.io/polars/py-polars/html/reference/ expressions/api/polars.Expr.str.strptime.html

- https://docs.rs/chrono/latest/chrono/format/strftime/index.html

Extracting substrings

Extracting substrings is a crucial component in string manipulation. It means deriving a portion of a string and using it as another column or as part of a transformation logic. Knowing how to extract substrings helps you clean, transform, and organize your data into a more useful format.

In this recipe, we'll cover how to extract substrings by slicing and regex.

How to do it...

Here's how you extract substrings from strings in Polars:

1. Use .str.slice() to extract a substring. There are two available parameters in this method: offset and length. The following example only specifies the offset:

```
df.select(
    'userName',
    pl.col('userName').str.slice(3).alias('4thCharAndAfter')
).head()
```

The preceding code will return the following output:

shape: (5, 2)

userName	4thCharAndAfter
str	str
"Eric Tie"	"c Tie"
"john alpha"	"n alpha"
"Sudhakar .S"	"hakar .S"
"SKGflorida@bellsouth.net DAVID S"	"florida@bellsouth.net DAVID S"
"Louann Stoker"	"ann Stoker"

Figure 6.11 – A new column with userName after the 4ᵗ character

2. You can specify how many characters you want to extract with `.str.slice()`:

```
df.select(
    'userName',
    pl.col('userName').str.slice(3,
5).alias('5CharsAfter4thChar')
).head()
```

The preceding code will return the following output:

shape: (5, 2)

userName	5CharsAfter4thChar
str	str
"Eric Tie"	"c Tie"
"john alpha"	"n alp"
"Sudhakar .S"	"hakar"
"SKGflorida@bellsouth.net DAVID S"	"flori"
"Louann Stoker"	"ann S"

Figure 6.12 – The five characters after the 4th character of userName

3. Specify the starting character with a negative number. This shows the second to last character:

```
df.select(
    'userName',
    pl.col('userName').str.slice(-2,
1).alias('TheLastToSecondChar')
).head()
```

The preceding code will return the following output:

shape: (5, 2)

userName	TheLastToSecondChar
str	str
"Eric Tie"	"i"
"john alpha"	"h"
"Sudhakar .S"	"."
"SKGflorida@bellsouth.net DAVID S"	" "
"Louann Stoker"	"e"

Figure 6.13 – The last to second character of userName

4. Use .str.extract() to derive a substring matched by the regex. The regex pattern gives you one or more alphabetical characters. Note that the .str.extract() method sets the capture group index to 1 by default:

```
df.select(
    'content',
    pl.col('content')
    .str.extract(r'([A-Za-z]+)')
    .alias('extract')
).head(5)
```

The preceding code will return the following output:

shape: (5, 2)

content	extract
str	str
"I cannot open the app anymore"	"I"
"I have been begging for a refund from this app fo..."	"I"
"Very costly for the premium version (approx India..."	"Very"
"Used to keep me organized, but all the 2020 UPDAT..."	"Used"
"Dan Birthday Oct 28"	"Dan"

Figure 6.14 – A new column created with str.extract

5. Specify the capture group to extract with .str.extract(). The whole regex that follows extracts the three alphabetical values followed by a space and one or more numbers:

```
df.select(
    'content',
    pl.col('content')
    .str.extract(r'([A-Za-z]{3}) ([0-9]+)', 0)
    .alias('extract whole matches specified'),
    pl.col('content')
```

```
        .str.extract(r'([A-Za-z]{3}) ([0-9]+)', 1)
        .alias('extract group 1 specified'),
    pl.col('content')
        .str.extract(r'([A-Za-z]{3}) ([0-9]+)', 2)
        .alias('extract group 2 specified')
).head(5)
```

The preceding code will return the following output:

shape: (5, 4)

content	extract whole matches specified	extract group 1 specified	extract group 2 specified
str	str	str	str
"I cannot open the app anymore"	null	null	null
"I have been begging for a refund from this app fo...	null	null	null
"Very costly for the premium version (approx India...	"ees 910"	"ees"	"910"
"Used to keep me organized, but all the 2020 UPDAT...	"the 2020"	"the"	"2020"
"Dan Birthday Oct 28"	"Oct 28"	"Oct"	"28"

Figure 6.15 – Extracted substrings in matched groupings

6. Extract all matches in the list with `.str.extract_all()`:

```
df.select(
    'content',
    pl.col('content')
    .str.extract(r'([A-Za-z]+)')
    .alias('extract'),
    pl.col('content')
    .str.extract_all(r'([A-Za-z]+)')
    .alias('extract_all')
).head(5)
```

The preceding code will return the following output:

shape: (5, 3)

content	extract	extract_all
str	str	list[str]
"I cannot open the app anymore"	"cannot"	["I", "cannot", ... "anymore"]
"I have been begging for a refund from this app fo...	"have"	["I", "have", ... "me"]
"Very costly for the premium version (approx India...	"costly"	["Very", "costly", ... "better"]
"Used to keep me organized, but all the 2020 UPDAT...	"to"	["Used", "to", ... "salary"]
"Dan Birthday Oct 28"	"Birthday"	["Dan", "Birthday", "Oct"]

Figure 6.16 – Extracting matches in the list with str.extract_all

7. Use `.str.extract_groups()` to extract all capture groups in struct:

```
df.select(
    'content',
    pl.col('content')
    .str.extract(r'([A-Za-z]{3}) ([0-9]+)', 0)
    .alias('extract'),
    pl.col('content')
    .str.extract_groups(r'([A-Za-z]{3}) ([0-9]+)')
    .alias('extract_groups')
).head()
```

The preceding code will return the following output:

shape: (5, 3)

content	extract	extract_groups
str	str	struct[2]
"I cannot open the app anymore"	null	{null,null}
"I have been begging for a refund from this app fo...	null	{null,null}
"Very costly for the premium version (approx India...	"ees 910"	{"ees","910"}
"Used to keep me organized, but all the 2020 UPDAT...	"the 2020"	{"the","2020"}
"Dan Birthday Oct 28"	"Oct 28"	{"Oct","28"}

Figure 6.17 – Matched groups in struct with str.extract_groups

How it works...

The `.str.slice()` method is simple to work with. You only need to specify the starting index and/or how many characters you want to extract.

Using regex is rather complex. Methods such as `.str.extract()`, `.str.extract_all()`, and `.str.extract_groups()` require you to learn some regex. I won't go into details about regex syntax, but I have listed a few resources under the *See also* section in this recipe.

`.str.extract()` extracts the target capture group. `.str.extract_all()` extracts all matches in list from the specified regex. `.str.extract_groups()` extracts all capture groups in struct.

We'll cover how to work with lists and structs in the next chapter.

There's more...

There are flags you can specify in Polars's regex. Flags can modify regex behavior. In the following example, I use the `case-insensitive` flag. I no longer need to specify lowercase alphabets:

```
df.select(
    'content',
```

```
    pl.col('content')
    .str.extract_all(r'(?i)n([A-Z]+)')
    .alias('extract_all')
).head()
```

The preceding code will return the following output:

shape: (5, 2)

content	extract_all
str	list[str]
"I cannot open the app anymore"	["I", "cannot", ... "anymore"]
"I have been begging for a refund from this app fo...	["I", "have", ... "me"]
"Very costly for the premium version (approx India...	["Very", "costly", ... "better"]
"Used to keep me organized, but all the 2020 UPDAT...	["Used", "to", ... "salary"]
"Dan Birthday Oct 28"	["Dan", "Birthday", "Oct"]

Figure 6.18 – A new column has been created using the case-insensitive flag

You can read more about flags here: `https://docs.rs/regex/latest/regex/#grouping-and-flags`.

See also

These resources are helpful in understanding the string methods introduced in this recipe in detail:

- `https://pola-rs.github.io/polars/user-guide/expressions/strings/`
- `https://pola-rs.github.io/polars/py-polars/html/reference/expressions/api/polars.Expr.str.slice.html`
- `https://pola-rs.github.io/polars/py-polars/html/reference/expressions/api/polars.Expr.str.extract.html`
- `https://pola-rs.github.io/polars/py-polars/html/reference/expressions/api/polars.Expr.str.extract_all.html`
- `https://docs.rs/regex/latest/regex/`
- `https://regex101.com/`

Cleaning strings

You may find your strings messy when you start working with your dataset for the first time. Inconsistent spacing, unexpected letter cases, odd spelling mistakes, and so on may be causing this. Your data may not be in a state in which you can start your analysis or the necessary data transformations. Knowing how to clean strings helps move past this roadblock.

In this recipe, we'll cover ways to clean strings using methods such as `.str.strip_chars()`, `.str.replace()`, and `.str.to_titlecase()`.

Getting ready

We're using a manually created dataset for this recipe. Run the following to create a DataFrame:

```
df = pl.DataFrame(
    {
        'text': [
            ' I aM a HUmAn... ',
            'it is NOT easy! ',
            ' WHY are You cool'
        ]
    }
)
```

How to do it...

Here are ways to clean strings:

1. Remove leading and trailing whitespaces with `.str.strip_chars()`:

    ```
    df.select(
        'text',
        pl.col('text')
        .str.strip_chars()
        .alias('stripped_text')
    )
    ```

    The preceding code will return the following output:

    shape: (3, 2)

text	stripped_text
str	str
" I aM a HUmAn. "	"I aM a HUmAn."
"it is NOT easy! "	"it is NOT easy!"
" WHY are You cool"	"WHY are You cool"

    Figure 6.19 – A DataFrame with a stripped text column

2. Replace matching substrings with `.str.replace()`.

    I. Replace the specified number of matches:

```
df.select(
    'text',
    pl.col('text')
    .str.replace('a', 'new_a', literal=True, n=1)
    .alias('replaced_text')
)
```

The preceding code will return the following output:

shape: (3, 2)

text	replaced_text
str	str
" I aM a HUmAn. "	" I new_aM a HUmAn. "
"it is NOT easy! "	"it is NOT enew_asy! "
" WHY are You cool"	" WHY new_are You cool"

Figure 6.20 – A DataFrame with replaced values

    II. Replace all the occurrences of the matched substring:

```
df.select(
    'text',
    pl.col('text')
    .str.replace_all('a', 'new_a', literal=True)
    .alias('replaced_all_text')
)
```

The preceding code will return the following output:

shape: (3, 2)

text	replaced_all_text
str	str
" I aM a HUmAn. "	" I new_aM new_a HUmAn. "
"it is NOT easy! "	"it is NOT enew_asy! "
" WHY are You cool"	" WHY new_are You cool"

Figure 6.21 – A DataFrame with all matched substrings replaced

3. Change the letter case of your strings.

    I. Change the letter case to title case:

```
df.select(
    'text',
    pl.col('text')
    .str.to_titlecase().alias('title_case')
)
```

The preceding code will return the following output:

shape: (3, 2)

text	title_case
str	str
" I aM a HUmAn. "	" I Am A Human. "
"it is NOT easy! "	"It Is Not Easy! "
" WHY are You cool"	" Why Are You Cool"

Figure 6.22 – A DataFrame with text changed to title case

    II. Change the letter case to lowercase:

```
df.select(
    'text',
    pl.col('text')
    .str.to_lowercase().alias('lower_case')
)
```

The preceding code will return the following output:

shape: (3, 2)

text	lower_case
str	str
" I aM a HUmAn. "	" i am a human. "
"it is NOT easy! "	"it is not easy! "
" WHY are You cool"	" why are you cool"

Figure 6.23 – A DataFrame with text changed to lowercase

    III. Change the letter case to uppercase:

```
df.select(
    'text',
```

```
      pl.col('text')
      .str.to_uppercase().alias('upper_case')
)
```

The preceding code will return the following output:

shape: (3, 2)

text	upper_case
str	str
" I aM a HUmAn. "	" I AM A HUMAN. "
"it is NOT easy! "	"IT IS NOT EASY! "
" WHY are You cool"	" WHY ARE YOU COOL"

Figure 6.24 – A DataFrame with text changed to uppercase

4. Add a certain character until the string reaches the specified length:

```
df.select(
    'text',
    pl.col('text')
    .str.pad_start(20, '~').alias('pad_start'),
    pl.col('text')
    .str.pad_end(20, '~').alias('pad_end'),
)
```

The preceding code will return the following output:

shape: (3, 3)

text	pad_start	pad_end
str	str	str
" I aM a HUmAn. "	"~~~ I aM a HUmAn. "	" I aM a HUmAn. ~~~"
"it is NOT easy! "	"~it is NOT easy! "	"it is NOT easy! ~"
" WHY are You cool"	"~~~ WHY are You cool"	" WHY are You cool~~~"

Figure 6.25 – A DataFrame with text padded until the string reaches a given length

How it works...

Most of the methods that have been introduced are self-explanatory. One thing to keep in mind is that in .str.replace(), you have the n parameter, which specifies how many matches to replace. Its default value is 1. Also, for both the .str.replace() and .str.replace_all() methods, you have the option to use regex.

See also

You can find additional information about the string methods used in this recipe in the following resources:

- `https://docs.pola.rs/api/python/stable/reference/series/api/polars.Series.str.strip_chars.html`
- `https://pola-rs.github.io/polars/py-polars/html/reference/series/api/polars.Series.str.replace.html`
- `https://docs.pola.rs/api/python/stable/reference/expressions/api/polars.Expr.str.to_titlecase.html`
- `https://docs.pola.rs/api/python/stable/reference/expressions/api/polars.Expr.str.pad_start.html`

Splitting strings into lists and structs

Understanding how to split strings into lists and structs is indispensable in data projects due to the pivotal role that splitting strings plays in data processing and feature engineering. Raw data often arrives in unstructured formats. The ability to split strings enables the extraction of meaningful information, facilitating data cleaning and transformation, particularly in data engineering and wrangling tasks.

List and struct data types will be the output when you split strings. We'll cover how to work with those data types, including the operations we can do on them, in *Chapter 7, Working with Nested Data Structures*.

In this recipe, we'll look at how we can split strings into lists and structs in Python Polars using the `.str.split()`, `.str.splitn()`, and `.str.split_exact()` methods.

Getting ready

We'll be using the Google Store review dataset we've been using throughout this chapter. Read in the dataset with the following code:

```
df = pl.read_csv('../data/google_store_reviews.csv')
```

How to do it...

Here's how to split strings into lists and structs.

1. Split strings into lists using `.str.split()`:

```
df.select(
    'content',
```

```
        pl.col('content').str.split(by=' ').alias('split')
).head()
```

The preceding code will return the following output:

shape: (5, 2)

content	split
str	list[str]
"I cannot open the app anymore"	["I", "cannot", ... "anymore"]
"I have been begging for a refund from this app fo...	["I", "have", ... "me"]
"Very costly for the premium version (approx India...	["Very", "costly", ... "better."]
"Used to keep me organized, but all the 2020 UPDAT...	["Used", "to", ... "😀😀😀"]
"Dan Birthday Oct 28"	["Dan", "Birthday", ... "28"]

Figure 6.26 – A new list column from the content column

2. Split strings into structs using `.str.splitn()` and `.str.split_exact()`:

```
df.select(
    'content',
    pl.col('content').str.splitn(by=' ', n=10).alias('splitn'),
    pl.col('content').str.split_exact(by=' ', n=10).
alias('split_exact')
).head()
```

The preceding code will return the following output:

shape: (5, 3)

content	splitn	split_exact
str	struct[10]	struct[11]
"I cannot open the app anymore"	{"I","cannot","open","the","app","anymore",null,null,null,null}	{"I","cannot","open","the","app","anymore",null,null,null,null,null}
"I have been begging for a refund from this app fo...	{"I","have","been","begging","for","a","refund","from","this","app for over a month and nobody is replying me"}	{"I","have","been","begging","for","a","refund","from","this","app","for"}
"Very costly for the premium version (approx India...	{"Very","costly","for","the","premium","version"," (approx","Indian","Rupees","910 per year). Better to download the premium version of this app from apkmos website and use it. Microsoft to do list app is far more better."}	{"Very","costly","for","the","premium","version"," (approx","Indian","Rupees","910","per"}
"Used to keep me organized, but all the 2020 UPDAT...	{"Used","to","keep","me","organized,","but","all","the","2020","UPDATES have made a mess of things !!! Y cudn't u leave well enuf alone ??? Guess ur techies feel the need to keep making changes to justify continuing to collect their salary !!! 😀😀😀"}	{"Used","to","keep","me","organized,","but","all","the","2020","UPDATES","have"}
"Dan Birthday Oct 28"	{"Dan","Birthday","Oct","28",null,null,null,null,null,null}	{"Dan","Birthday","Oct","28",null,null,null,null,null,null,null}

Figure 6.27 – New struct columns are derived from the content column

How it works...

The `.str.split()` method splits the string into each item of a list by the given delimiter.

On the other hand, the `.str.splitn()` and `.str.split_exact()` methods split the string into each item of a struct by the given delimiter.

`.str.splitn()` splits the string into n fields in a struct. When n is greater than the number of resulting fields in the struct, then null values are added. When n is less than the number of resulting fields in the struct, then the last substring will contain the remainder of the string.

`.str.split_exact()` also splits the string into a struct, but it's different in that the resulting struct contains $n+1$ fields. Also, when n is less than the number of resulting fields in the struct, only $n+1$ number of fields are added in the struct.

See also

These resources will help you learn more about splitting strings in Python Polars:

- https://pola-rs.github.io/polars/py-polars/html/reference/expressions/api/polars.Expr.str.split.html

- https://pola-rs.github.io/polars/py-polars/html/reference/expressions/api/polars.Expr.str.splitn.html

- https://docs.pola.rs/api/python/stable/reference/expressions/api/polars.Expr.str.split_exact.html

Concatenating and combining strings

Concatenating or combining strings is another way to derive meaningful information from your text data. The ability to concatenate and combine strings is a fundamental skill in the data domain. It facilitates data preparation, feature engineering, and the integration of disparate datasets, contributing to the overall success of any data project.

In this recipe, we'll cover a few ways to concatenate strings in Python Polars.

Getting ready

In this recipe, we'll use a DataFrame that has been created manually:

```python
df = pl.DataFrame(
    {
        'colA': ['a', 'b', 'c', 'd'],
        'colB': ['aa', 'bb', 'cc', 'dd']
    }
)
```

How to do it...

Here are a few methods you can use:

1. Use + to concatenate strings.

    Using a literal or constant value:

    ```
    df.select(
        pl.all(),
        (pl.col('colB') + ' new').alias('newColB')
    )
    ```

    The preceding code will return the following output:

    shape: (4, 3)

colA	colB	newColB
str	str	str
"a"	"aa"	"aa new"
"b"	"bb"	"bb new"
"c"	"cc"	"cc new"
"d"	"dd"	"dd new"

    Figure 6.28 – The new column is created by combining a column and a string

2. Concatenate strings from multiple columns into a single column.

    I.   Using the + symbol:

    ```
    df.select(
        pl.all(),
        (pl.col('colA') + pl.col('colB')).alias('colC')
    )
    ```

    The preceding code will return the following output:

    shape: (4, 3)

colA	colB	colC
str	str	str
"a"	"aa"	"aaa"
"b"	"bb"	"bbb"
"c"	"cc"	"ccc"
"d"	"dd"	"ddd"

    Figure 6.29 – The new column is created by combining multiple columns

II.   Using `pl.concat_str()`:

```
df.select(
    pl.all(),
    pl.concat_str(
        pl.lit(100)+3,
        pl.lit(' '),
        pl.col('colA'),
        pl.col('colB'),
        separator='::'
    ).alias('newCol')
)
```

The preceding code will return the following output:

shape: (4, 3)

colA	colB	newCol
str	str	str
"a"	"aa"	"103:: ::a::aa"
"b"	"bb"	"103:: ::b::bb"
"c"	"cc"	"103:: ::c::cc"
"d"	"dd"	"103:: ::d::dd"

Figure 6.30 – The new column is created by combining multiple columns and literals

3. Use `.str.join()` to combine all of the strings of a column into one string. In other words, it concatenates strings vertically across rows:

```
df.select(
    pl.all(),
    pl.col('colA').str.join(delimiter=',
').alias('concatenatedColA')
)
```

The preceding code will return the following output:

shape: (4, 3)

colA	colB	concatenatedColA
str	str	str
"a"	"aa"	"a, b, c, d"
"b"	"bb"	"a, b, c, d"
"c"	"cc"	"a, b, c, d"
"d"	"dd"	"a, b, c, d"

Figure 6.31 – The new column is created by combining strings vertically across rows

How it works...

To combine strings from multiple columns into a single column, you can use + or `pl.concat_str()` to concatenate them together. The difference is that the latter option allows you to specify the separator in the parameter. You can also pass in the numeric data type without explicitly casting them to string data type. When using the + operator, you'd first need to cast the numeric data type value to `string/Utf8`. That is one of the reasons why you may want to use `pl.concat_str()` rather than using + for concatenation. A few other reasons include the fact that `pl.concat_str()` is more descriptive and allows you to add separators automatically.

The `.str.join()` method lets you specify the delimiter and decide whether to ignore nulls with the `ignore_nulls` parameter. When it's set to `False`, it returns null as the output if a column contains a null value.

See also

Please refer to these additional resources to learn more about concatenating strings in Python Polars:

- `https://docs.pola.rs/api/python/stable/reference/expressions/api/polars.concat_str.html`
- `https://docs.pola.rs/api/python/stable/reference/expressions/api/polars.Expr.str.join.html`

7

Working with Nested Data Structures

Proficiency in working with nested data types during data wrangling is essential for effectively handling the complexity of real-world datasets. As datasets increasingly manifest intricate structures and relationships, understanding how to navigate and manipulate nested data becomes crucial.

The ability to work with nested data is fundamental for advanced data analysis, machine learning, and data wrangling. This enables the creation of sophisticated features and facilitates compatibility with modern data interchange formats and complex business logic. In essence, mastering the intricacies of nested data types enhances data professional's capacity for sophisticated data analysis and modeling in the field of data.

Python Polars provides built-in functionalities to work with nested data structures such as `List`, `Array`, `Struct`, and `Object`. `List` is a data type in which each row represents a list of elements with the same data type but different lengths. On the other hand, `Array` is a data type in which each row represents a fixed-size array of elements with the same data type. `Struct` is a data type in which each row represents a collection of columns. It gives you the flexibility to implement transformation logic on multiple columns at once. Finally, `Object` is similar to `List` and `Array` in that it holds a list of elements. It's different because it contains different data types in the list.

The aim of this chapter is to introduce you to ways to work with those nested data structures. We'll do that using practical examples that teach you exactly what you need to get started.

In this chapter, the key things we're going to cover are list and struct operations. The chapter includes the following recipes:

- Creating lists
- Aggregating elements in lists
- Accessing and selecting elements in lists

- Applying logic to each element in lists
- Working with structs and JSON data

Technical requirements

You can download the datasets and code in the GitHub repository:

- Data: `https://github.com/PacktPublishing/Polars-Cookbook/tree/main/data`
- Code: `https://github.com/PacktPublishing/Polars-Cookbook/tree/main/Chapter07`

It is assumed that you have installed the Polars library in your Python environment:

```
>>> pip install polars
```

It is also assumed that you have imported it into your code:

```
import polars as pl
```

We'll be using the Trending YouTube Video Statistics dataset (`https://github.com/PacktPublishing/Polars-Cookbook/blob/main/data/us_videos.csv`) throughout this chapter. Read the data before proceeding to the next steps:

```
df = pl.read_csv('../data/us_videos.csv', try_parse_dates=True)
```

Also, look at what the data looks like in the dataset:

```
df.glimpse(max_items_per_column=2)
```

The preceding code will produce the following output:

```
Rows: 40949
Columns: 16
$ video_id                                      <str> '2kyS6SvSYSE', '1ZAPwfrtAFY'
$ trending_date                                 <str> '17.14.11', '17.14.11'
$ title                                         <str> 'WE WANT TO TALK ABOUT OUR MARRIAGE', 'The Trump Presidency: Last Week Tonight with John Oliver (HBO)'
$ channel_title                                 <str> 'CaseyNeistat', 'LastWeekTonight'
$ category_id                                   <i64> 22, 24
$ publish_time          <datetime[μs, UTC]> 2017-11-13 17:13:01+00:00, 2017-11-13 07:30:00+00:00
$ tags                                          <str> 'SHANtell martin', 'last week tonight trump presidency|last week tonight donald trump|john oliver trump|donald trump'
$ views                                         <i64> 748374, 2418783
$ likes                                         <i64> 57527, 97185
$ dislikes                                      <i64> 2966, 6146
$ comment_count                                 <i64> 15954, 12703
$ thumbnail_link                                <str> 'https://i.ytimg.com/vi/2kyS6SvSYSE/default.jpg', 'https://i.ytimg.com/vi/1ZAPwfrtAFY/default.jpg'
$ comments_disabled                            <bool> False, False
$ ratings_disabled                             <bool> False, False
$ video_error_or_removed                       <bool> False, False
$ description                                   <str> "SHANTELL'S CHANNEL - https://www.youtube.com/shantellmartin\\nCANDICE - https://www.lovebilly.com\\n\\nfilmed this vi
```

Figure 7.1 – An overview of the DataFrame

Let's fix the data type for the `trending_date` column:

```
df = df.with_columns(
pl.col('trending_date').str.strptime(pl.Date, format='%y.%d.%m')
)
df.select('trending_date').dtypes[0]
```

The preceding code will return the following output:

```
>> Date
```

Creating lists

There are only two occasions when you will have columns of the `List` data type. One is when Polars recognizes the `List` data type upon reading data. Another is when you create a `list` column in your code. Whether it's splitting a string into a list of strings or combining values from multiple columns, creating lists is the start of your complex analysis involving nested data structures.

In this recipe, we'll cover how to create lists by splitting strings, grouping by columns, and combining multiple values into lists.

How to do it...

Here's how to create lists:

- Create lists from strings using `.str.split()` to split strings into lists:

```
df.select(
    'tags',
    pl.col('tags').str.split('|').alias('tags in list')
).head()
```

The preceding code will return the following output:

shape: (5, 2)

tags	tags in list
str	list[str]
"SHANtell marti...	["SHANtell martin"]
"last week toni...	["last week tonight trump presidency", "last week tonight donald trump", ... "donald trump"]
"racist superma...	["racist superman", "rudy", ... " Lele Pons"]
"rhett and link...	["rhett and link", "gmm", ... "challenge"]
"ryan|higa|higa...	["ryan", "higa", ... "fail"]

Figure 7.2 – The first five rows in the DataFrame with a list column

- Creating lists by using a group by operation. Let's group videos by their trending date:

```
(
    df
    .group_by('trending_date')
    .agg(pl.col('video_id'))
    .sort('trending_date', descending=True)
).head()
```

The preceding code will return the following output:

shape: (5, 2)

trending_date	video_id
date	list[str]
2018-06-14	["-QPdRfqTnt4", "gPHVLxm8U-0", … "ooyjaVdt-jA"]
2018-06-13	["FchkqXEg0qs", "uHRwMmwbFnA", … "Q5KmA3Xbmqo"]
2018-06-12	["PPWDwBrUNyY", "rAH8qm5oQHg", … "6S9c5nnDd_s"]
2018-06-11	["0bXCbVGb04A", "L4pkD78oKSo", … "r-3iathMo7o"]
2018-06-10	["L4pkD78oKSo", "ZFwyIDNpgFc", … "r-3iathMo7o"]

Figure 7.3 – YouTube videos by trending date

- Creating lists using `pl.concat_list()`:

```
df.select(
    pl.concat_list(
        pl.col('views'),
        pl.col('likes'),
        pl.col('dislikes'),
        pl.col('comment_count')
    ).alias('engagement')
).head()
```

The preceding code will return the following output:

shape: (5, 1)

engagement
list[i64]
[748374, 57527, … 15954]
[2418783, 97185, … 12703]
[3191434, 146033, … 8181]
[343168, 10172, … 2146]
[2095731, 132235, … 17518]

Figure 7.4 – A new list column, created by concatenating existing columns

How it works...

Turning strings into lists is a common operation in data projects. Oftentimes, your datasets will not be organized in the way you want. They may also not be ready for analysis and further engineering right away. You may find some data is stored as strings when it should be stored in the list for your use case. That's when the `.str.split()` method comes in handy.

The `.group_by()` method can be used to create lists when you don't specify aggregations in the `.agg()` method. It aggregates the values spread across rows for each group into lists.

`pl.concat_list()` combines columns horizontally into a list column. It's very much like string concatenation, but it instead outputs the result into a list.

There's more...

You can even create lists of lists, or nested lists:

```
df = pl.DataFrame({
    'nested_list': [
        [
            [1,2,3], [4,5,6],
            [7,8,9], [10,11,12]
        ],
        [
            [1,2,3], [4,5,6],
            [7,8,9], [10,11,12]
        ]
    ]
})
```

The DataFrame will look like the following:

shape: (2, 1)

nested_list
list[list[i64]]
[[1, 2, 3], [4, 5, 6], ... [10, 11, 12]]
[[1, 2, 3], [4, 5, 6], ... [10, 11, 12]]

Figure 7.5 – A DataFrame with a column of the nested list data type

If you add an element whose data type is different from other elements and set the `strict` parameter to `False`, or else your code will raise an error. Let's change the very first element in the first list to strings. This will make the data type of list elements from integer to string:

```
df = pl.DataFrame({
    'nested_list': [
        [
            ['a',2,3], [4,5,6],
            [7,8,9], [10,11,12]
        ],
        [
            [1,2,3], [4,5,6],
            [7,8,9], [10,11,12]
        ]
    ]
}, strict=False)
```

The DataFrame will look like the following:

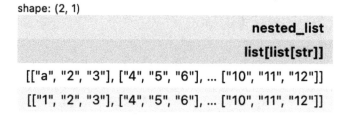

Figure 7.6 – A DataFrame with an object data type column

Although what I've just demonstrated is possible to do, that doesn't mean you should do it. Using mixed data types will give you more headaches than utility.

See also

These additional resources will help you understand more about the methods and expressions used in the recipe:

- https://docs.pola.rs/py-polars/html/reference/expressions/api/polars.Expr.str.split.html

- https://docs.pola.rs/py-polars/html/reference/dataframe/api/polars.DataFrame.group_by.html

- https://docs.pola.rs/py-polars/html/reference/expressions/api/
  polars.concat_list.html

Aggregating elements in lists

Aggregations are a way to summarize data. Whether your list is populated by numeric values or strings, you can aggregate elements in the list in Python Polars.

In this recipe, we'll cover how to aggregate elements in lists using common aggregations such as `min` and `max`, as well as how to turn a list of strings into a string.

How to do it...

Here are ways to aggregate data in lists:

1. Let's create a DataFrame with columns of the `list` data type, where we have YouTube video statistics per trending date:

```
agg_df = (
    df
    .group_by('trending_date')
    .agg(
        'views',
        'likes',
        'dislikes',
        'comment_count'
    )
    .sort('trending_date', descending=True)
)
agg_df.head()
```

The preceding code will return the following output:

shape: (5, 5)

trending_date	views	likes	dislikes	comment_count
date	list[i64]	list[i64]	list[i64]	list[i64]
2018-06-14	[4427381, 5829270, … 10306119]	[96391, 87323, … 357079]	[5508, 3668, … 212976]	[12726, 11933, … 144795]
2018-06-13	[3238183, 470844, … 7839668]	[61841, 13922, … 352352]	[3708, 402, … 5871]	[0, 4843, … 46624]
2018-06-12	[3483553, 6173038, … 13619534]	[23725, 90478, … 347100]	[3145, 3877, … 6923]	[462, 7726, … 19977]
2018-06-11	[2341772, 846887, … 6995168]	[140374, 5758, … 143678]	[2951, 850, … 10925]	[33760, 1539, … 17444]
2018-06-10	[594004, 1504564, … 6980540]	[4470, 32430, … 143452]	[520, 4316, … 10911]	[1195, 6105, … 17429]

Figure 7.7 – YouTube video statistics per trending date

2. To calculate basic aggregations over list columns, use the following:

```
(
    agg_df
    .select(
        'trending_date',
        pl.col('views').list.min().alias('views_min'),
        pl.col('likes').list.max().alias('likes_max'),
        pl.col('dislikes').list.mean().alias('dislikes_mean'),
        pl.col('comment_count').list.sum().alias('comment_sum'),
    )
    .sort('trending_date', descending=True)
).head()
```

The preceding code will return the following output:

shape: (5, 5)

trending_date	views_min	likes_max	dislikes_mean	comment_sum
date	i64	i64	f64	i64
2018-06-14	189038	2032463	8472.94	3289433
2018-06-13	175754	2021395	8541.23	3379632
2018-06-12	161782	2004753	8353.685	3352328
2018-06-11	136643	1981942	8478.095	3320709
2018-06-10	116841	1967904	8405.22	3351105

Figure 7.8 – Basic aggregations of YouTube statistics per trending date

3. You can also aggregate strings in lists using the `.list.join()` method. Turning lists of channel titles into strings is possible with the following code:

```
(
    df
    .group_by('trending_date')
    .agg(pl.col('channel_title'))
    .with_columns(
        pl.col('channel_title').list.join(':')
    )
    .sort('trending_date', descending=True)
).head()
```

The preceding code will return the following output:

shape: (5, 2)

trending_date	channel_title
date	str
2018-06-14	"Disney Movie T...
2018-06-13	"Nintendo:GameX...
2018-06-12	"gameslice:Clas...
2018-06-11	"The Game Theor...
2018-06-10	"NBC Sports:Ant...

Figure 7.9 – Channel titles as strings per trending date

How it works...

The Polars has the `list` namespace in the Expressions API that contains many useful methods to work with lists, including basic aggregations such as `.list.min()`, `.list.max()`, and `.list.mean()`. This `list` namespace not only gives these useful methods and ways to work with lists, but it also gives clarity on what the code is about to do next.

The `.list.join()` method helps turn strings in a list into a single string with a specified separator. The inner data type needs to be `string/Utf8` for it to work. Think of this method as the inverse of `.str.split()`, which we covered in the previous chapter.

There's more...

You can count the number of elements in lists using the `.list.len()` method. Here's how to do that:

```
(
    agg_df
    .select(
        'trending_date',
        pl.col('views').list.len().alias('item_cnt')
    )
).head()
```

The preceding code will return the following output:

shape: (5, 2)

trending_date	item_cnt
date	u32
2018-03-09	200
2018-02-22	199
2018-02-25	200
2018-02-28	199
2018-03-03	199

Figure 7.10 – Item count in lists per trending date

See also

These resources are helpful in learning more about aggregating elements in lists:

- `https://docs.pola.rs/py-polars/html/reference/expressions/list.html`
- `https://docs.pola.rs/user-guide/expressions/lists/`

Accessing and selecting elements in lists

Accessing and selecting elements in lists is a common operation that is useful for further transforming and analyzing your data. Being able to freely access and select list elements extends your ability to wrangle data more efficiently. The good news is that Polars provides many ways to access and select list elements.

In this recipe, I'll cover ways to access and select elements in lists using list methods such as `.list.first()`, `.list.head()`, `.list.get()`, `.list.slice()`, `.list.contains()`, and so on.

Getting ready

We'll prepare the DataFrame so that it has trending dates per channel. Notice that we are using the `.list.sort()` method to sort elements in the list:

```
trending_dates_by_channel = (
    df
    .group_by('channel_title')
    .agg('trending_date')
```

```
        .with_columns(pl.col('trending_date').list.sort())
)
trending_dates_by_channel.head()
```

The preceding code will return the following output:

shape: (5, 2)

channel_title	trending_date
str	list[date]
"Drew Lynch"	[2018-01-24, 2018-01-25, ... 2018-02-12]
"FIFATV"	[2017-12-02, 2017-12-03, ... 2017-12-07]
"SmarterEveryDa...	[2017-12-29, 2017-12-30, ... 2018-03-03]
"GingerPale"	[2018-04-19, 2018-04-20, ... 2018-05-14]
"Linkin Park"	[2017-11-28, 2017-11-29, ... 2018-01-05]

Figure 7.11 – List of trending dates per channel

How to do it...

Here are ways to access and select elements in lists:

1. You can get the first and last element in the list using the `.list.first()` and `.list.last()` methods:

    ```
    trending_dates_by_channel.with_columns(
        pl.col('trending_date').list.first().alias('first_trending_
    date'),
        pl.col('trending_date').list.last().alias('last_trending_
    date')
    ).head()
    ```

    The preceding code will return the following output:

shape: (5, 4)

channel_title	trending_date	first_trending_date	last_trending_date
str	list[date]	date	date
"Drew Lynch"	[2018-01-24, 2018-01-25, ... 2018-02-12]	2018-01-24	2018-02-12
"FIFATV"	[2017-12-02, 2017-12-03, ... 2017-12-07]	2017-12-02	2017-12-07
"SmarterEveryDa...	[2017-12-29, 2017-12-30, ... 2018-03-03]	2017-12-29	2018-03-03
"GingerPale"	[2018-04-19, 2018-04-20, ... 2018-05-14]	2018-04-19	2018-05-14
"Linkin Park"	[2017-11-28, 2017-11-29, ... 2018-01-05]	2017-11-28	2018-01-05

Figure 7.12 – The first and last trending dates in the list

2. You can get a particular element in lists with the `.list.get()` method, specifying the index of the element:

```
trending_dates_by_channel.with_columns(
    pl.col('trending_date').list.get(7, null_on_oob=True).
alias('8th_element')

).head()
```

The preceding code will return the following output:

shape: (5, 3)

channel_title	trending_date	8th_element
str	list[date]	date
"Drew Lynch"	[2018-01-24, 2018-01-25, … 2018-02-12]	2018-01-25
"FIFATV"	[2017-12-02, 2017-12-03, … 2017-12-07]	2017-12-03
"SmarterEveryDa…	[2017-12-29, 2017-12-30, … 2018-03-03]	2017-12-30
"GingerPale"	[2018-04-19, 2018-04-20, … 2018-05-14]	2018-04-20
"Linkin Park"	[2017-11-28, 2017-11-29, … 2018-01-05]	2017-11-29

Figure 7.13 – Showing the eighth element in the list

3. Use `.list.head()` and `.list.tail()` to show the first and last *N* elements in lists:

```
trending_dates_by_channel.with_columns(
    pl.col('trending_date').list.head().alias('first_5'),
    pl.col('trending_date').list.tail(10).alias('last_10')
).head()
```

The preceding code will return the following output:

shape: (5, 4)

channel_title	trending_date	first_5	last_10
str	list[date]	list[date]	list[date]
"Drew Lynch"	[2018-01-24, 2018-01-25, … 2018-02-12]	[2018-01-24, 2018-01-25, … 2018-01-28]	[2018-01-25, 2018-01-26, … 2018-02-12]
"FIFATV"	[2017-12-02, 2017-12-03, … 2017-12-07]	[2017-12-02, 2017-12-03, … 2017-12-06]	[2017-12-02, 2017-12-03, … 2017-12-07]
"SmarterEveryDa…	[2017-12-29, 2017-12-30, … 2018-03-03]	[2017-12-29, 2017-12-30, … 2018-01-02]	[2018-02-06, 2018-02-07, … 2018-03-03]
"GingerPale"	[2018-04-19, 2018-04-20, … 2018-05-14]	[2018-04-19, 2018-04-20, … 2018-04-23]	[2018-04-26, 2018-04-27, … 2018-05-14]
"Linkin Park"	[2017-11-28, 2017-11-29, … 2018-01-05]	[2017-11-28, 2017-11-29, … 2017-12-07]	[2017-12-27, 2017-12-28, … 2018-01-05]

Figure 7.14 – Showing the first and last N elements in the list

4. You can get the top three elements in lists using `.list.sort()` and `.list.head()` in combination:

```
trending_dates_by_channel.with_columns(
    pl.col('trending_date')
    .list.sort(descending=True)
    .list.head(3)
    .alias('3_most_recent_dates')
).head()
```

The preceding code will return the following output:

shape: (5, 3)

channel_title	trending_date	3_most_recent_dates
str	list[date]	list[date]
"Drew Lynch"	[2018-01-24, 2018-01-25, … 2018-02-12]	[2018-02-12, 2018-02-11, 2018-02-10]
"FIFATV"	[2017-12-02, 2017-12-03, … 2017-12-07]	[2017-12-07, 2017-12-06, 2017-12-05]
"SmarterEveryDa…	[2017-12-29, 2017-12-30, … 2018-03-03]	[2018-03-03, 2018-03-02, 2018-02-13]
"GingerPale"	[2018-04-19, 2018-04-20, … 2018-05-14]	[2018-05-14, 2018-05-04, 2018-05-03]
"Linkin Park"	[2017-11-28, 2017-11-29, … 2018-01-05]	[2018-01-05, 2018-01-04, 2018-01-03]

Figure 7.15 – Showing the 3 most recent dates

5. You can get sublist using the `.list.slice()` method:

```
trending_dates_by_channel.select(
    'trending_date',
    pl.col('trending_date').list.slice(0, 2).alias('first_2_dates'),
    pl.col('trending_date').list.slice(-3, 1).alias('3rd_date_to_last'),
    pl.col('trending_date').list.slice(7).alias('from_8th_date_to_end')
).head()
```

The preceding code will return the following output:

shape: (5, 4)

trending_date	first_2_dates	3rd_date_to_last	from_8th_date_to_end
list[date]	list[date]	list[date]	list[date]
[2018-03-01, 2018-03-02, … 2018-06-14]	[2018-03-01, 2018-03-02]	[2018-06-12]	[2018-03-08, 2018-03-09, … 2018-06-14]
[2018-05-15, 2018-05-15]	[2018-05-15, 2018-05-15]	[]	[]
[2017-11-20, 2017-11-21, … 2018-03-08]	[2017-11-20, 2017-11-21]	[2018-03-06]	[2017-12-01, 2017-12-02, … 2018-03-08]
[2017-11-14, 2017-11-15, … 2017-11-19]	[2017-11-14, 2017-11-15]	[2017-11-17]	[]
[2017-11-18, 2017-11-19, … 2018-05-08]	[2017-11-18, 2017-11-19]	[2018-05-06]	[2017-11-25, 2017-11-26, … 2018-05-08]

Figure 7.16 – Getting selected elements in lists with the .list.slice() method

6. You can get sublists with `.list.gather()`, specifying multiple indices:

```
trending_dates_by_channel.select(
    'trending_date',
    pl.col('trending_date').list.gather([0, -1]).alias('first_
and_last'),
    pl.col('trending_date').list.gather([0, 0, 0, 0]).
alias('first_repeated'),
    pl.col('trending_date').list.gather([0, 10], null_on_
oob=True).alias('first_and_10th_or_null')
).head()
```

The preceding code will return the following output:

shape: (5, 4)

trending_date	first_and_last	first_repeated	first_and_10th_or_null
list[date]	list[date]	list[date]	list[date]
[2018-01-24, 2018-01-25, … 2018-02-12]	[2018-01-24, 2018-02-12]	[2018-01-24, 2018-01-24, … 2018-01-24]	[2018-01-24, 2018-02-12]
[2017-12-02, 2017-12-03, … 2017-12-07]	[2017-12-02, 2017-12-07]	[2017-12-02, 2017-12-02, … 2017-12-02]	[2017-12-02, null]
[2017-12-29, 2017-12-30, … 2018-03-03]	[2017-12-29, 2018-03-03]	[2017-12-29, 2017-12-29, … 2017-12-29]	[2017-12-29, 2018-02-09]
[2018-04-19, 2018-04-20, … 2018-05-14]	[2018-04-19, 2018-05-14]	[2018-04-19, 2018-04-19, … 2018-04-19]	[2018-04-19, 2018-04-29]
[2017-11-28, 2017-11-29, … 2018-01-05]	[2017-11-28, 2018-01-05]	[2017-11-28, 2017-11-28, … 2017-11-28]	[2017-11-28, 2017-12-24]

Figure 7.17 – Selected elements in lists with the .list.gather() method

How it works...

The `.list.get()` method helps you extract a particular element from a list with an index specified. The resulting value will have the data type as in the original list. The `null_on_oob` parameter needs to be set to True if you want your code to return Null if an index is out of bounds. Otherwise, your code will raise an error.

The `.list.slice()` method works similarly to how `.str.slice()`. It involves specifying the offset and length.

The `.list.gather()` method lets you specify multiple indices to select elements from a list. You don't get to specify the length, but you specify multiple elements by its indices. You can also set the `null_on_oob` parameter so that if the specified index is out of bound, then it returns null.

There's more...

There are a few more useful methods when accessing and selecting elements in lists. The `.list.unique()` method helps you extract only the unique elements in the list:

```
(
    df
    .group_by('trending_date')
    .agg('category_id')
    .with_columns(pl.col('category_id').list.sort())
    .with_columns(
        pl.col('category_id'),
        pl.col('category_id').list.len().alias('category_id_cnt'),
        pl.col('category_id').list.unique().alias('category_id_
unique'),
        pl.col('category_id').list.unique().list.len().
alias('category_id_unique_cnt')
    )
).head()
```

The preceding code will return the following output:

shape: (5, 5)

trending_date	category_id	category_id_cnt	category_id_unique	category_id_unique_cnt
date	list[i64]	u32	list[i64]	u32
2018-02-16	[1, 1, ... 29]	200	[1, 2, ... 29]	15
2018-03-12	[1, 1, ... 28]	199	[1, 10, ... 28]	13
2018-02-04	[1, 1, ... 29]	196	[1, 2, ... 29]	14
2018-03-09	[1, 1, ... 28]	200	[1, 10, ... 28]	13
2018-02-28	[1, 1, ... 28]	199	[1, 2, ... 28]	14

Figure 7.18 – Extracting unique elements in lists with .list.unique()

The `.list.sample()` method lets you sample data from the list:

```
trending_dates_by_channel.with_columns(
    pl.col('trending_date')
    .list.sample(n=3, with_replacement=True, seed=0)
    .alias('samples')
).head()
```

The preceding code will return the following output:

shape: (5, 3)

channel_title	trending_date	samples
str	list[date]	list[date]
"Drew Lynch"	[2018-01-24, 2018-01-25, ... 2018-02-12]	[2018-01-28, 2018-01-28, 2018-02-12]
"FIFATV"	[2017-12-02, 2017-12-03, ... 2017-12-07]	[2017-12-04, 2017-12-04, 2017-12-07]
"SmarterEveryDa...	[2017-12-29, 2017-12-30, ... 2018-03-03]	[2018-02-06, 2018-02-06, 2018-03-03]
"GingerPale"	[2018-04-19, 2018-04-20, ... 2018-05-14]	[2018-04-26, 2018-04-26, 2018-05-14]
"Linkin Park"	[2017-11-28, 2017-11-29, ... 2018-01-05]	[2017-12-24, 2017-12-24, 2018-01-05]

Figure 7.19 – A new column with sample values from trending_date

Although it is typically a sensible choice to do sampling without replacement, I set the with_replacement parameter to True in the preceding example. This is largely for demonstration purposes, as well as the fact that some of the trending_date lists contain less than three elements, which would make it necessary to sample with replacement. Otherwise, the code would throw an error.

You can specify other parameters for the .list.sample() method. Please refer to those parameters in the Polars documentation: https://docs.pola.rs/py-polars/html/reference/expressions/api/polars.Expr.list.sample.html.

See also

Please refer to these resources to learn more about working with nested data in Polars:

- https://docs.pola.rs/user-guide/expressions/lists/
- https://docs.pola.rs/user-guide/expressions/structs/
- https://docs.pola.rs/py-polars/html/reference/expressions/list.html
- https://docs.pola.rs/py-polars/html/reference/expressions/struct.html

Applying logic to each element in lists

One of Polars's strengths is the ability to work well with nested structures. It not only helps with aggregations, as well as selecting and accessing elements in lists but also helps you apply simple to complex logic to each element in a list. This gives you the flexibility and power to transform your data in an efficient manner utilizing Polars's optimizations.

In this recipe, we'll cover how you can apply transformation logic to each element in a list.

Getting ready

We'll be using a pre-aggregated DataFrame for this recipe. We will aggregate views, likes, and channel titles by trending date:

```
agg_df = (
    df
    .group_by('trending_date')
    .agg('views', 'likes', 'channel_title')
)
agg_df.head()
```

The preceding code will return the following output:

shape: (5, 3)

trending_date	views	channel_title
date	list[i64]	list[str]
2018-02-13	[266874, 953801, ... 736748]	["NBC Sports", "The King of Random", ... "SmarterEveryDay"]
2018-03-06	[4852889, 3141764, ... 3021666]	["Dude Perfect", "Disney Movie Trailers", ... "SMTOWN"]
2018-02-19	[832910, 470720, ... 173997]	["What's Inside?", "Strictly Dumpling", ... "Food Wishes"]
2018-02-22	[2841864, 1234544, ... 111619]	["TheEllenShow", "Unbox Therapy", ... "Physics Girl"]
2018-02-04	[209539, 1497616, ... 180238]	["This Is Us", "FBE", ... "Sports Illustrated"]

Figure 7.20 – A DataFrame with views and channel_title aggregated into lists

How to do it...

Here is how to apply logic to each element in lists, complete with examples:

1. Functions such as `pl.element()`, `pl.first()`, and `pl.last()` help select each element in lists. The following code is just accessing each element in the list:

```
(
    agg_df
    .select(
        'views',
        pl.col('views').list.eval(pl.element()).alias('pl.
element'),
        pl.col('views').list.eval(pl.first()).alias('pl.first'),
        pl.col('views').list.eval(pl.last()).alias('pl.last'),
        pl.col('views').list.eval(pl.col('')).alias('pl.col')
    )
).head()
```

The preceding code will return the following output:

shape: (5, 5)

views	pl.element	pl.first	pl.last	pl.col
list[i64]	list[i64]	list[i64]	list[i64]	list[i64]
[10096037, 744691, … 302660]	[10096037, 744691, … 302660]	[10096037, 744691, … 302660]	[10096037, 744691, … 302660]	[10096037, 744691, … 302660]
[266874, 953801, … 736748]	[266874, 953801, … 736748]	[266874, 953801, … 736748]	[266874, 953801, … 736748]	[266874, 953801, … 736748]
[2311760, 1397595, … 2926283]	[2311760, 1397595, … 2926283]	[2311760, 1397595, … 2926283]	[2311760, 1397595, … 2926283]	[2311760, 1397595, … 2926283]
[209539, 1497616, … 180238]	[209539, 1497616, … 180238]	[209539, 1497616, … 180238]	[209539, 1497616, … 180238]	[209539, 1497616, … 180238]
[10582444, 1894717, … 39595]	[10582444, 1894717, … 39595]	[10582444, 1894717, … 39595]	[10582444, 1894717, … 39595]	[10582444, 1894717, … 39595]

Figure 7.21 – Different ways to specify each element in lists

2. Apply a string operation on each list element for the `channel_title` column, changing strings to uppercase and showing the first two elements:

```
channel_titles_df = (
    agg_df
    .select(
        pl.col("channel_title").list.head(2),
        pl.col("channel_title")
        .list.eval(pl.element().str.to_uppercase())
        .list.head(2)
        .alias("channel_title_upper")
    )
)
channel_titles_df.head()
```

The preceding code will return the following output:

shape: (5, 2)

channel_title	channel_title_upper
list[str]	list[str]
["NBC Sports", "The King of Random"]	["NBC SPORTS", "THE KING OF RANDOM"]
["Dude Perfect", "Disney Movie Trailers"]	["DUDE PERFECT", "DISNEY MOVIE TRAILERS"]
["What's Inside?", "Strictly Dumpling"]	["WHAT'S INSIDE?", "STRICTLY DUMPLING"]
["TheEllenShow", "Unbox Therapy"]	["THEELLENSHOW", "UNBOX THERAPY"]
["This Is Us", "FBE"]	["THIS IS US", "FBE"]

Figure 7.22 – Values in the channel title converted to uppercase

3. Filter elements in lists where we only keep elements containing A using the DataFrame we just created:

```
(
    channel_titles_df
    .with_columns(
```

```
            pl.col('channel_title_upper')
            .list.eval(
                pl.element().filter(
                    pl.element().str.contains('A', literal=True)
                )
            )
        )
    ).head()
```

The preceding code will return the following output:

shape: (5, 2)

channel_title	channel_title_upper
list[str]	list[str]
["Nintendo", "CamilaCabelloVEVO"]	["CAMILACABELLOVEVO"]
["The Late Show with Stephen Colbert", "The Slow Mo Guys"]	["THE LATE SHOW WITH STEPHEN COLBERT"]
["Dude Perfect", "Disney Movie Trailers"]	["DISNEY MOVIE TRAILERS"]
["carrieunderwoodVEVO", "nigahiga"]	["CARRIEUNDERWOODVEVO", "NIGAHIGA"]
["Disney·Pixar", "Mr. Kate"]	["DISNEY·PIXAR", "MR. KATE"]

Figure 7.23 – Showing channel titles containing the letter A

4. Add a new `views_rank` column that ranks the number of views:

```
(
    agg_df
    .select(
        'trending_date',
        'views',
        pl.col('views')
        .list.eval(pl.element().rank('dense', descending=True)
        .alias('views_rank')
    )
).head()
```

The preceding code will return the following output:

shape: (5, 3)

trending_date	views	views_rank
date	list[i64]	list[u32]
2018-02-13	[266874, 953801, … 736748]	[133, 66, … 79]
2018-03-06	[4852889, 3141764, … 3021666]	[10, 20, … 22]
2018-02-19	[832910, 470720, … 173997]	[55, 78, … 137]
2018-02-22	[2841864, 1234544, … 111619]	[18, 49, … 158]
2018-02-04	[209539, 1497616, … 180238]	[122, 29, … 131]

Figure 7.24 – Adding a column ranking views in the list

Let's filter the whole DataFrame so that it only contains the top three most views.

5. First, we will turn the query used in the preceding step into a DataFrame so that we can work off of it for this recipe:

```
views_rank_df = (
    agg_df
    .select(
        'trending_date',
        'views',
        pl.col('views')
        .list.eval(pl.element().rank('dense', descending=True))
        .alias('views_rank'),
    )
)
```

Use the .explode() method to turn list elements into rows. Then group the data again by the views and views_rank columns to turn those filtered rows into lists:

```
top3_views_df = (
    views_rank_df
    .explode('views', 'views_rank')
    .filter(pl.col('views_rank')<=3)
    .group_by('trending_date')
    .agg(pl.all())
)
top3_views_df.head()
```

The preceding code will return the following output:

shape: (5, 3)

trending_date	views	views_rank
date	list[i64]	list[u32]
2018-02-10	[51243149, 25193150, 20015015]	[1, 2, 3]
2018-03-02	[8258921, 10695328, 17190739]	[3, 2, 1]
2017-12-09	[75969469, 35696409, 28860037]	[1, 2, 3]
2018-01-12	[18871288, 7055170, 57951412]	[2, 3, 1]
2017-11-27	[24628712, 34043133, 20752341]	[2, 1, 3]

Figure 7.25 – A DataFrame containing the top three views with ranks

6. Let's see how the output will look like if we only apply the `.explode()` method without grouping back the DataFrame:

```
(
    views_rank_df
    .explode('views', 'views_rank')
).head()
```

The preceding code will return the following output:

shape: (5, 3)

trending_date	views	views_rank
date	i64	u32
2018-02-13	266874	133
2018-02-13	953801	66
2018-02-13	638618	85
2018-02-13	319444	122
2018-02-13	3780988	23

Figure 7.26 – The views and views_rank columns expanded into rows

7. Add a column that calculates the difference between the maximum value (views with first rank) and each value in each list:

```
(
    top3_views_df
    .select(
        'views_rank',
        'views',
        pl.col('views')
```

```
            .list.eval(pl.element().max() - pl.element())
            .alias('diff from the most views')
        )
    ).head()
```

The preceding code will return the following output:

shape: (5, 3)

views_rank	views	diff from the most views
list[u32]	list[i64]	list[i64]
[2, 3, 1]	[28475675, 23575103, 38070298]	[9594623, 14495195, 0]
[3, 2, 1]	[30173222, 43738208, 60962220]	[30788998, 17224012, 0]
[1, 3, 2]	[40801423, 14882505, 18052108]	[0, 25918918, 22749315]
[2, 3, 1]	[15295839, 12180415, 52404970]	[37109131, 40224555, 0]
[1, 3, 2]	[35830721, 17983051, 18241671]	[0, 17847670, 17589050]

Figure 7.27 – A new column calculating the between the maximum value and each value in each list

How it works...

The .list.eval() method is the key to applying logic to each element in lists. It essentially allows you to access each list element and run any Polars expression against them.

There are a few ways in which you can specify list elements within .list.eval(), as demonstrated in *step 1*. Those include pl.element(), pl.first(), and so forth. Once you have defined such an expression, you can start chaining other expressions like you do on any other columns.

> **Tip**
> My recommendation and preferred approach are to use pl.element() within the .list.eval() method to select list elements. This is because it's the most descriptive option. It tells you that the code is evaluating list elements.

Step 5 demonstrates how you can apply your transformation on a list column based on another list column. Filtering list elements on one column doesn't automatically filter another list column that has the same length. In this case, turning lists into rows is a good way to accomplish what you want. This assumes that the column you want to apply the logic to and the column you're basing your logic on have the same number of elements. Otherwise, Polars wouldn't turn list elements into rows for multiple list columns using the .explode() method.

You might want to just reference the column within .list.eval(). However, as of the writing of this book, you cannot refer to another column in the method (you can learn more about this on GitHub at https://github.com/pola-rs/polars/issues/9948). You can refer to the elements of the list that the .list.eval() method is operating on, as you saw in *step 6*.

> **Tip**
>
> In *step 5*, we turned lists into rows to filter the whole DataFrame. Then we grouped the DataFrame back to what it was: the `views` and `views_rank` columns aggregated into lists per `trending_date` by simply using the `.group_by()` and `.agg()` methods. The `.implode()` method also helps aggregate values into a list, but it's different from the combination of `.group_by()` and `.agg()` in that the `.implode()` method doesn't change the granularity of the DataFrame. On the other hand, using `.group_by()` and `.agg()` consolidates values into groups, thus changing the granularity of the DataFrame.

There's more...

Polars provides set operations over list columns. A set operation operates on two list columns and obtains the result as per the operation. As for the writing of this book, Polars provides list methods for calculating the intersection, union, difference, and symmetric difference of two sets. Here's an example of how we can use such list methods:

```
(
    top3_views_df
    .with_columns(
        pl.col('views_rank').list.slice(0, 2).alias('views_rank_1'),
        pl.col('views_rank').list.slice(-2, 2).alias('views_rank_2')
    )
    .select(
        'views_rank_1',
        'views_rank_2',
        pl.col('views_rank_1').list.set_intersection('views_rank_2')
        .alias('intersection'),
        pl.col('views_rank_1').list.set_union('views_rank_2')
        .alias('union'),
        pl.col('views_rank_1').list.set_difference('views_rank_2')
        .alias('difference'),
        pl.col('views_rank_1').list.set_symmetric_difference('views_
rank_2')
        .alias('symmetric_difference'),
    )
).head()
```

The preceding code will return the following output:

shape: (5, 6)

views_rank_1	views_rank_2	intersection	union	difference	symmetric_difference
list[u32]	list[u32]	list[u32]	list[u32]	list[u32]	list[u32]
[1, 3]	[3, 2]	[3]	[1, 3, 2]	[1]	[1, 2]
[3, 2]	[2, 1]	[2]	[3, 2, 1]	[3]	[3, 1]
[1, 3]	[3, 2]	[3]	[1, 3, 2]	[1]	[1, 2]
[2, 1]	[1, 3]	[1]	[2, 1, 3]	[2]	[2, 3]
[2, 1]	[1, 3]	[1]	[2, 1, 3]	[2]	[2, 3]

Figure 7.28 – New columns with set operations applied

See also

Please refer to these resources to learn more about applying logic to elements in lists:

- `https://docs.pola.rs/user-guide/expressions/lists/`
- `https://docs.pola.rs/py-polars/html/reference/dataframe/api/polars.DataFrame.explode.html#polars.DataFrame.explode`
- `https://docs.pola.rs/py-polars/html/reference/expressions/api/polars.Expr.implode.html`
- `https://docs.pola.rs/py-polars/html/reference/expressions/api/polars.Expr.list.eval.html#polars.Expr.list.eval`
- `https://docs.pola.rs/py-polars/html/reference/expressions/api/polars.element.html`
- `https://docs.pola.rs/py-polars/html/reference/expressions/list.html`

Working with structs and JSON data

Struct is another nested data type in Polars. It represents a collection of columns. It helps you pack multiple columns into a `struct` column on which you can apply operations. One good use case is to apply your transformation logic to the unique combinations of multiple columns.

JSON data is often stored as structs in a DataFrame. You can also work with JSON as strings. Polars has built-in methods to encode and decode JSON data between string and struct.

In this recipe, we'll look at how to work with structs and JSON data.

Getting ready

We'll use a Google Analytics dataset (which can be found at `https://github.com/PacktPublishing/Polars-Cookbook/blob/main/data/ga_20170801.json`) for this recipe. Read a JSON file into a DataFrame with the following code:

```
df = pl.read_json('../data/ga_20170801.json')
cols = ['visitId', 'date', 'totals', 'trafficSource',
'customDimensions', 'channelGrouping']

df = df.select(cols)
df.head()
```

The preceding code will return the following output:

shape: (5, 6)

visitId	date	totals	trafficSource	customDimensions	channelGrouping
str	str	struct[13]	struct[9]	list[struct[2]]	str
"1501591568"	"20170801"	{"1","1","1",null,"1",null,null,"1",null,null,null,null,"1"}	{null,"(not set)","","(direct)","(none)",null,null, {null,null,null,null,null,null,"not available in demo dataset",null,null,null,null,null},null,null}	[]	"Organic Search...
"1501589647"	"20170801"	{"1","1","1",null,"1",null,null,null,null,null,null,null,"1"}	{"/analytics/web/","",("not set)","analytics.google.com","referral",null,null, {null,null,null,null,null,null,"not available in demo dataset",null,null,null,null,null},null,null}	[{"4","APAC"}]	"Referral"
"1501616621"	"20170801"	{"1","1","1",null,"1",null,null,"1",null,null,null,null,"1"}	{"/analytics/web/","",("not set)","analytics.google.com","referral",null,null, {null,null,null,null,null,null,"not available in demo dataset",null,null,null,null,null},null,null}	[{"4","EMEA"}]	"Referral"
"1501601200"	"20170801"	{"1","1","1",null,"1",null,null,"1",null,null,null,null,"1"}	{"/analytics/web/","",("not set)","analytics.google.com","referral",null,null, {null,null,null,null,null,null,"not available in demo dataset",null,null,null,null,null},null,null}	[{"4","North America"}]	"Referral"
"1501615525"	"20170801"	{"1","1","1",null,"1",null,null,"1",null,null,null,null,"1"}	{"/analytics/web/","",("not set)","adwords.google.com","referral",null,null, {null,null,null,null,null,null,"not available in demo dataset",null,null,null,null,null},null,null}	[{"4","North America"}]	"Referral"

Figure 7.29 – The first five rows in the Google Analytics dataset

Let's look closely at a few columns whose values are stored as JSON:

shape: (5, 3)

totals	trafficSource	customDimensions
struct[13]	struct[9]	list[struct[2]]
{"1","1","1",null,"1",null,null,"1",null,null,null,null,"1"}	{null,"(not set)","(direct)","(none)",null,null, {null,null,null,null,null,null,"not available in demo dataset",null,null,null,null,null},null,null}	[]
{"1","1","1",null,"1",null,null,null,null,null,null,null,"1"}	{"/analytics/web/","(not set)","analytics.google.com","referral",null,null, {null,null,null,null,null,null,"not available in demo dataset",null,null,null,null,null},null,null}	[{"4","APAC"}]
{"1","1","1",null,"1",null,null,"1",null,null,null,null,"1"}	{"/analytics/web/","(not set)","analytics.google.com","referral",null,null, {null,null,null,null,null,null,"not available in demo dataset",null,null,null,null,null},null,null}	[{"4","EMEA"}]
{"1","1","1",null,"1",null,null,"1",null,null,null,null,"1"}	{"/analytics/web/","(not set)","analytics.google.com","referral",null,null, {null,null,null,null,null,null,"not available in demo dataset",null,null,null,null,null},null,null}	[{"4","North America"}]
{"1","1","1",null,"1",null,null,"1",null,null,null,null,"1"}	{"/analytics/web/","(not set)","adwords.google.com","referral",null,null, {null,null,null,null,null,null,"not available in demo dataset",null,null,null,null,null},null,null}	[{"4","North America"}]

Figure 7.30 – The struct data type holds JSON data

The `totals` and `trafficSource` columns are read as `struct` data types. The `customDimensions` column is read as a list of structs. Note that a column of the `struct` data type gives you the number of elements in the `struct` when using the `.head()` method. Let's look at the data source file to see how these are stored in their original format.

Here's how the `totals` column is stored in its original format:

```
"totals": {
    "visits": "1",
    "hits": "1",
    "pageviews": "1",
    "timeOnSite": null,
    "bounces": "1",
    "transactions": null,
    "transactionRevenue": null,
    "newVisits": null,
    "screenviews": null,
    "uniqueScreenviews": null,
    "timeOnScreen": null,
    "totalTransactionRevenue": null,
    "sessionQualityDim": "1"
},
```

Figure 7.31 – The totals column in its original format

Here's how the `trafficSource` column is stored in its original format:

```
"trafficSource": {
  "referralPath": "/analytics/web/",
  "campaign": "(not set)",
  "source": "analytics.google.com",
  "medium": "referral",
  "keyword": null,
  "adContent": null,
  "adwordsClickInfo": {
    "campaignId": null,
    "adGroupId": null,
    "creativeId": null,
    "criteriaId": null,
    "page": null,
    "slot": null,
    "criteriaParameters": "not available in demo dataset",
    "gclId": null,
    "customerId": null,
    "adNetworkType": null,
    "targetingCriteria": null,
    "isVideoAd": null
  },
  "isTrueDirect": null,
  "campaignCode": null
},
```

Figure 7.32 – The trafficSource column in its original format

Here's how the `customDimensions` column is stored in its original format:

```
"customDimensions": [{
  "index": "4",
  "value": "APAC"
}],
```

Figure 7.33 – The customDimensions column in its original format

We can see that the `totals` and `trafficSource` columns are stored as JSON and that the `customDimensions` column stores data as a list of JSON. Looking at *Figure 7.30*, you will notice that Polars reads data as it is stored in the source file.

> **Note**
>
> Notice that the value of each key-value pair is translated into the value in the struct. You may wonder where the *key* to the key-value pair went. We don't see it right now. However, when we expand the struct, we'll see those key names, which were stored in the DataFrame after all. We'll see this in action in the steps discussed in the following section.

How to do it...

Let's look at some of the essential operations for structs and JSON data:

1. Create a `struct` column from multiple columns:

```
df = df.with_columns(
    pl.struct('visitId', 'date', 'channelGrouping').
alias('structFromCols')
)


(
    df
    .select(
        'visitId',
        'date',
        'channelGrouping',
        pl.struct('visitId', 'date', 'channelGrouping').
alias('structFromCols')
    )
).head()
```

The preceding code will return the following output:

shape: (5, 4)

visitId	date	channelGrouping	structFromCols
str	str	str	struct[3]
"1501591568"	"20170801"	"Organic Search...	{"1501591568","20170801","Organic Search"}
"1501589647"	"20170801"	"Referral"	{"1501589647","20170801","Referral"}
"1501616621"	"20170801"	"Referral"	{"1501616621","20170801","Referral"}
"1501601200"	"20170801"	"Referral"	{"1501601200","20170801","Referral"}
"1501615525"	"20170801"	"Referral"	{"1501615525","20170801","Referral"}

Figure 7.34 – The customDimensions column in its original format

2. Create a `struct` column from a list column:

```
(
    df
    .group_by('channelGrouping')
    .agg(
        'visitId',
```

```
        pl.col('visitId').len().alias('numVisits')
    )
    .sort('numVisits')
    .with_columns(
        pl.col('visitId').list.to_struct().alias('struct_from_
list')
    )
)
```

The preceding code will return the following output:

shape: (7, 4)

channelGrouping	visitId	numVisits	struct_from_list
str	list[str]	u32	struct[12]
"Display"	["1501651856", "1501625928", … "1501638116"]	12	{"1501651856","1501625928","1501611633","1501625068","1501612878","1501616158","1501607332","1501622703","1501621231","1501649570","1501647417","1501638116"}
"Paid Search"	["1501610896", "1501644116", … "1501613648"]	20	{"1501610896","1501644116","1501574187","1501614708","1501626398","1501636143","1501617844","1501628199","1501618328","1501624893","1501650118","1501580936"}
"Affiliates"	["1501604627", "1501572101", … "1501635918"]	29	{"1501604627","1501572101","1501638418","1501589505","1501635523","1501588687","1501602677","1501656633","1501589444","1501654572","1501588961","1501588902"}
"Referral"	["1501589647", "1501616621", … "1501607798"]	106	{"1501589647","1501616621","1501601200","1501615525","1501589650","1501573710","1501613382","1501630140","1501656976","1501602227","1501620300","1501611288"}
"Social"	["1501590147", "1501655923", … "1501652602"]	136	{"1501590147","1501655923","1601640054","1501696419","1501591307","1501618949","1501649584","1501579329","1501585068","1501618027","1501653304","1501614595"}
"Direct"	["1501586309", "1501587435", … "1501610792"]	163	{"1501586309","1501587435","1501653660","1501608816","1501611913","1501584277","1501578373","1501587465","1501621325","1501655032","1501622827","1501575169"}
"Organic Search…	["1501591568", "1501583103", … "1501626964"]	534	{"1501591568","1501583103","1501631547","1501599064","1501585229","1501639903","1501576309","1501573981","1501618526","1501576968","1501599268","1501596177"}

Figure 7.35 – Adding a struct column, which was created from a list column

3. Unpack a `struct` column back into multiple columns with the `.unnest()` method:

```
(
    df
    .select(
        'structFromCols',
        pl.col('structFromCols').
alias('structFromColsToBeUnpacked')
    )
    .unnest('structFromColsToBeUnpacked')
).head()
```

The preceding code will return the following output:

shape: (5, 4)

structFromCols	visitId	date	channelGrouping
struct[3]	str	str	str
{"1501591568","20170801","Organic Search"}	"1501591568"	"20170801"	"Organic Search...
{"1501589647","20170801","Referral"}	"1501589647"	"20170801"	"Referral"
{"1501616621","20170801","Referral"}	"1501616621"	"20170801"	"Referral"
{"1501601200","20170801","Referral"}	"1501601200"	"20170801"	"Referral"
{"1501615525","20170801","Referral"}	"1501615525"	"20170801"	"Referral"

Figure 7.36 – A struct column unpacked into columns

Notice that Polars retains the original column names. However, what about the JSON data that we read as `struct` columns, such as the `totals` and `trafficSource` columns, that only shows us the values, not the keys?

4. Let's unpack the `trafficSource` column and see what this gives you:

```
(
    df
    .select(
        pl.col('trafficSource')
    )
    .unnest('trafficSource')
).head()
```

The preceding code will return the following output:

shape: (5, 9)

referralPath	campaign	source	medium	keyword	adContent	adwordsClickInfo	isTrueDirect	campaignCode
str	str	str	str	str	null	struct[12]	str	null
null	"(not set)"	"(direct)"	"(none)"	null	null	{null,null,null,null,null,null,"not available in demo dataset",null,null,null,null,null}	null	null
"/analytics/web...	"(not set)"	"analytics.goog...	"referral"	null	null	{null,null,null,null,null,null,"not available in demo dataset",null,null,null,null,null}	null	null
"/analytics/web...	"(not set)"	"analytics.goog...	"referral"	null	null	{null,null,null,null,null,null,"not available in demo dataset",null,null,null,null,null}	null	null
"/analytics/web...	"(not set)"	"analytics.goog...	"referral"	null	null	{null,null,null,null,null,null,"not available in demo dataset",null,null,null,null,null}	null	null
"/analytics/web...	"(not set)"	"adwords.google...	"referral"	null	null	{null,null,null,null,null,null,"not available in demo dataset",null,null,null,null,null}	null	null

Figure 7.37 – The trafficSource column Unpacked

Notice that you can see all those keys as column names for the values you didn't see earlier in *Figure 7.32*. Polars indeed had it all, but simply didn't show it in structs unless you unpacked the values.

Also, look at the `adwordsClickInfo` column. It's a `struct` column, meaning that Polars can have nested structures within nested structures.

5. Rename the keys of a `struct` column with the `.struct.rename_fields()` method. You can see those names when you unpack the `struct` column like you did in the preceding step:

```
(
    df
    .select(
        'structFromCols',
        pl.col('structFromCols').struct.rename_fields(['a', 'b',
'c']).alias('renamedStructToBeUnpacked')

    )
    .unnest('renamedStructToBeUnpacked')
).head()
```

The preceding code will return the following output:

shape: (5, 4)

structFromCols	a	b	c
struct[3]	str	str	str
{"1501591568","20170801","Organic Search"}	"1501591568"	"20170801"	"Organic Search...
{"1501589647","20170801","Referral"}	"1501589647"	"20170801"	"Referral"
{"1501616621","20170801","Referral"}	"1501616621"	"20170801"	"Referral"
{"1501601200","20170801","Referral"}	"1501601200"	"20170801"	"Referral"
{"1501615525","20170801","Referral"}	"1501615525"	"20170801"	"Referral"

Figure 7.38 – Unpacking a struct column with its fields renamed

6. Extract values of a `struct` column using the `.struct.field()` method:

```
(
    df
    .select(
        'structFromCols',
        pl.col('structFromCols').struct.field('channelGrouping')
    )
).head()
```

The preceding code will return the following output:

shape: (5, 2)

structFromCols	channelGrouping
struct[3]	str
{"1501591568","20170801","Organic Search"}	"Organic Search...
{"1501589647","20170801","Referral"}	"Referral"
{"1501616621","20170801","Referral"}	"Referral"
{"1501601200","20170801","Referral"}	"Referral"
{"1501615525","20170801","Referral"}	"Referral"

Figure 7.39 – Unpacking one column from a struct column

7. Get the unique combinations of multiple columns with `pl.struct()` and `.unique()`:

```
(
    df
    .select(
        pl.struct(
            pl.col('channelGrouping'),
            pl.col('trafficSource').struct.field('source')
        )
        .unique()
        .alias('channelAndSource')
    )
    .unnest('channelAndSource')
    .sort('channelGrouping', 'source')
)
```

The preceding code will return the following output:

shape: (28, 2)

channelGrouping	source
str	str
"Affiliates"	"Partners"
"Direct"	"(direct)"
"Display"	"(direct)"
"Display"	"dfa"
"Organic Search...	"(direct)"
"Organic Search...	"ask"
"Organic Search...	"baidu"
"Paid Search"	"(direct)"
"Referral"	"(direct)"
"Referral"	"adwords.google...
"Referral"	"analytics.goog...
"Referral"	"blog.golang.or...
...	...
"Referral"	"ph.search.yaho...
"Referral"	"productforums....
"Referral"	"qiita.com"
"Referral"	"sashihara.jp"
"Referral"	"sites.google.c...
"Referral"	"support.google...
"Social"	"facebook.com"
"Social"	"groups.google....
"Social"	"l.facebook.com...
"Social"	"m.facebook.com...
"Social"	"quora.com"
"Social"	"youtube.com"

Figure 7.40 – The unique combination of channelGrouping and source

How to do it...

The pl.struct() function helps create a struct column from multiple columns.

The .list.to_struct() method converts a list column into a struct column. In *step 2*, the resulting struct column only has 12 fields or elements even though the maximum number of elements in the original list column is 543. That's because the .list.to_struct() method assigns the number of fields to be equivalent to the length of the first list that does not have a length of 0 by default. In our case, the first list that does not have a length of 0 is the list of length 12. That's why the resulting struct column had a length of 12 fields. You can set the n_field_strategy parameter to max_width so that the number of fields will be the max length of all lists, which is 543. Null values would be added to supplement if a list doesn't meet the max length.

The `struct` data type not only helps you work with nested data formats such as JSON but it also helps compact your columns and apply logic to multiple columns with ease.

There's more...

JSON data could also be stored in formats such as `string` and `binary`, depending on how your original data is processed and stored. Especially with the `binary` data type, there is not much you can do until you have the data into a different data type to further transform data.

The `.cast()` method is useful to convert the data type of the column to something such as `string` or `struct` that allows you to utilize methods to work with JSON data efficiently.

Also, you can convert back and forth between `struct` and `string` as appropriate. Here's an example of converting a `struct` column to a string column using the `.struct.json_encode()` and converting it back to `struct` using the `.str.json_decode()` method:

```
total_struct_to_str_expr = pl.col('totals').struct.json_encode()
(
    df
    .select(
        total_struct_to_str_expr.alias('total_str'),
        total_struct_to_str_expr
        .str.json_decode()
        .alias('total_struct')
    )
).head()
```

The preceding code will return the following output:

shape: (5, 2)

total_str	total_struct
str	struct[13]
"{"visits":"1",...	{"1","1","1",null,"1",null,null,"1",null,null,null,null,"1"}
"{"visits":"1",...	{"1","1","1",null,"1",null,null,null,null,null,null,null,"1"}
"{"visits":"1",...	{"1","1","1",null,"1",null,null,"1",null,null,null,null,"1"}
"{"visits":"1",...	{"1","1","1",null,"1",null,null,"1",null,null,null,null,"1"}
"{"visits":"1",...	{"1","1","1",null,"1",null,null,"1",null,null,null,null,"1"}

Figure 7.41 – The totals columns have been converted back and forth

Notice that in the preceding code, I stored an expression in a variable and was able to use it in a context. That's totally legal and a good way to save lines of code. https://docs.pola.rs/py-polars/html/reference/expressions/struct.html

See also

To learn more about working with structs and JSON data, please refer to the following resources:

- `https://docs.pola.rs/user-guide/expressions/structs/`
- `https://docs.pola.rs/py-polars/html/reference/expressions/struct.html`
- `https://docs.pola.rs/py-polars/html/reference/expressions/api/polars.Expr.str.json_decode.html#polars.Expr.str.json_decode`
- `https://docs.pola.rs/py-polars/html/reference/dataframe/api/polars.DataFrame.unnest.html#polars.DataFrame.unnest`

8

Reshaping and Tidying Data

When analyzing and transforming data, it's not always the case that the data is in the best shape possible for your purpose. Your data may not be as clean or organized as you hoped. It may not even have the necessary column attributes for your analysis. This is where the concept of reshaping and tidying data comes into play. Reshaping means transforming data so that it suits a particular analysis you're trying to conduct. Tidying data refers to the process of organizing and structuring data in a clean and consistent way so that it is easy to work with and analyze. The process of reshaping and tidying data involves operations such as pivoting, unpivoting, stacking, and joining.

In this chapter, we're going to cover the following main topics:

- Turning columns into rows
- Turning rows into columns
- Joining DataFrames
- Concatenating DataFrames
- Other techniques for reshaping data

Technical requirements

You can download the dataset and code from the GitHub repository:

- Dataset: `https://github.com/PacktPublishing/Polars-Cookbook/tree/main/data`
- Code: `https://github.com/PacktPublishing/Polars-Cookbook/tree/main/Chapter08`

It is assumed that you have installed the Polars library in your Python environment:

```
>>> pip install polars
```

And that you imported it in your code:

```
import polars as pl
```

We'll be using the international student demographic dataset in this chapter. There are a few CSV files in the dataset we'll be using, but `academic.csv` will be used throughout the chapter. Read it into a DataFrame with the following code:

```
df = pl.read_csv('../data/academic.csv')
```

Clean the data by renaming a column, casting data types, and keeping only the recent years of data:

```
from polars import selectors as cs
df = (
    df
    .select(
        pl.col('year').alias('academic_year'),
        cs.numeric().cast(pl.Int64)
    )
    .filter(pl.col('academic_year').str.slice(0,4).cast(pl.
Int32)>=2018)
)
df.head()
```

The preceding code will return the following output:

shape: (5, 7)

academic_year	students	us_students	undergraduate	graduate	non_degree	opt
str	i64	i64	i64	i64	i64	i64
"2018/19"	1095299	19828000	431930	377943	62341	223085
"2019/20"	1075496	19720000	419321	374435	58201	223539
"2020/21"	914095	19744000	359787	329272	21151	203885
"2021/22"	948519	20327000	344532	385097	34131	184759
"2022/23"	1057188	18961280	347602	467027	43766	198793

Figure 8.1 – The first five rows of the DataFrame

Turning columns into rows

Unpivoting is a common operation done when reshaping or tidying your data. It helps turn and compact columns into rows. In other words, the unpivot operation converts your data from a wide format to a long format. This operation is also known as unpivot.

Both wide format and long format have their use cases. Some client tools prefer wide format while some prefer long format.

In this recipe, we'll cover how to turn columns into rows using the unpivot operation.

How to do it...

Here's how to turn columns into rows:

1. Use the `.unpivot()` method to turn columns into rows:

```python
long_df = df.unpivot(
    index='academic_year',
        on=[
            'students',
            'us_students',
            'undergraduate',
            'graduate',
            'non_degree',
            'opt'
        ],
    variable_name='student_type',
    value_name='count'
)
long_df.head()
```

The preceding code will return the following output:

shape: (5, 3)

academic_year	student_type	count
str	str	i64
"2018/19"	"students"	1095299
"2019/20"	"students"	1075496
"2020/21"	"students"	914095
"2021/22"	"students"	948519
"2022/23"	"students"	1057188

Figure 8.2 – The DataFrame in a long format

2. Let's check the `student_type` column to make sure it contains the values we expect:

```python
long_df.select('student_type').unique()
```

The preceding code will return the following output:

shape: (5, 1)

student_type
str
"opt"
"non_degree"
"undergraduate"
"us_students"
"graduate"

Figure 8.3 – Unique values in the student_type column

3. Utilize **selectors** in Polars to select columns at once for the same unpivot operation. This time, we won't change the variable and value names:

```
df.unpivot(
    index='academic_year',
    on=cs.numeric()
).head()
```

The preceding code will return the following output:

shape: (5, 3)

academic_year	variable	value
str	str	i64
"2018/19"	"students"	1095299
"2019/20"	"students"	1075496
"2020/21"	"students"	914095
"2021/22"	"students"	948519
"2022/23"	"students"	1057188

Figure 8.4 – The DataFrame in a long format without renaming new columns

4. Apply the unpivot operation in a LazyFrame.

Let's say you had a LazyFrame instead of a DataFrame:

```
lf = df.lazy()
```

You can still apply the unpivot operation in a LazyFrame:

```
(
    lf
    .unpivot(
        index='academic_year',
        on=cs.numeric(),
        variable_name='student_type',
        value_name='count'
    )
    .collect()
    .head()
)
```

The preceding code will return the same output as *Figure 8.4*.

How it works...

Using the `.unpivot()` method is straightforward. You specify the columns to keep and the columns that get turned into rows.

In *step 3*, we didn't specify the variable and value names. In that case, Polars automatically names those columns `variable` and `value`, respectively.

Also, as you know, many DataFrame methods are available in a LazyFrame; the `.unpivot()` method is one of them, which is great news.

See also

Feel free to refer to these resources to learn more about turning columns into rows:

- https://docs.pola.rs/api/python/stable/reference/dataframe/api/polars.DataFrame.unpivot.html

- https://docs.pola.rs/api/python/stable/reference/dataframe/api/polars.DataFrame.pivot.html

Turning rows into columns

The pivot operation works the opposite way from how the unpivot and unpivot operations work. It turns rows into columns or converts your data from a long to a wide format. A wide format is often easier to read and understand. Also, when conducting statistical analyses such as regression analysis or building a predictive model, having each variable as individual columns is more convenient.

In this recipe, we'll look at a few ways to turn rows into columns or convert a long format to a wide format.

Getting ready

In this recipe, we'll start with the DataFrame in the long format as created in the last recipe. If you haven't gone through that recipe, here's the code you need to execute to create the DataFrame:

```
from polars import selectors as cs

long_df = (
    pl.read_csv('../data/academic.csv')
    .select(
        pl.col('year').alias('academic_year'),
        cs.numeric().cast(pl.Int64)
    )
    .filter(
        pl.col('academic_year')
        .str.slice(0,4).cast(pl.Int32)>=2018
    )
    .unpivot(
        index='academic_year',
        on=[
            'students',
            'us_students',
            'undergraduate',
            'graduate',
            'non_degree',
            'opt'
        ],
        variable_name='student_type',
        value_name='count'
    )
)
long_df.head()
```

The preceding code will return the following output, which is the same as *Figure 8.4*:

shape: (5, 3)

academic_year	student_type	count
str	str	i64
"2018/19"	"students"	1095299
"2019/20"	"students"	1075496
"2020/21"	"students"	914095
"2021/22"	"students"	948519
"2022/23"	"students"	1057188

Figure 8.5 – The DataFrame in a long format

How to do it...

Here's how to turn rows into columns:

1. Use the `.pivot()` method to expand values in a column into individual columns:

```
(
    long_df
    .pivot(
        index='academic_year',
        values='count',
        on='student_type'
    )
)
```

The preceding code will return the following output:

shape: (5, 7)

academic_year	students	us_students	undergraduate	graduate	non_degree	opt
str	i64	i64	i64	i64	i64	i64
"2018/19"	1095299	19828000	431930	377943	62341	223085
"2019/20"	1075496	19720000	419321	374435	58201	223539
"2020/21"	914095	19744000	359787	329272	21151	203885
"2021/22"	948519	20327000	344532	385097	34131	184759
"2022/23"	1057188	18961280	347602	467027	43766	198793

Figure 8.6 – The DataFrame in a wide format

2. Use the combination of the `.group_by()` and `.agg()` methods to convert a long format to a wide format:

```
(
    long_df
    .group_by('academic_year', maintain_order=True)
    .agg(
        pl.col('count').filter(pl.col('student_
type')=='students').sum().alias('students'),
        pl.col('count').filter(pl.col('student_type')=='us_
students').sum().alias('us_students'),
        pl.col('count').filter(pl.col('student_
type')=='undergraduate').sum().alias('undergraduate'),
        pl.col('count').filter(pl.col('student_
type')=='graduate').sum().alias('graduate'),
        pl.col('count').filter(pl.col('student_type')=='non_
degree').sum().alias('non_degree'),
```

```
                pl.col('count').filter(pl.col('student_type')=='opt').
        sum().alias('opt')
            )
        )
```

The preceding code will return the same output as *Figure 8.6*.

You can also make those aggregated columns dynamic. That would be useful when you have many values in the column you're unpacking. See the following code for the logic to generate columns dynamically:

```
student_types = [
    col for col in long_df.select('student_type').unique().to_
series().to_list()
]

agg_cols = [
    (
        pl.col('count')
        .filter(pl.col('student_type')==stu_type)
        .sum()
        .alias(stu_type)
    )
    for stu_type in student_types
]


(
    long_df
    .group_by('academic_year', maintain_order=True)
    .agg(agg_cols)
)
```

The preceding code will return the same output as *Figure 8.6*. Note that the order of the columns may be different.

3. Use the `.unstack()` method to achieve a similar result as seen in *step 1* and *step 2*:

```
long_df.unstack(step=5, columns='count', how='vertical')
```

The preceding code will return the following output:

shape: (5, 6)

count_0	count_1	count_2	count_3	count_4	count_5
i64	i64	i64	i64	i64	i64
1095299	19828000	431930	377943	62341	223085
1075496	19720000	419321	374435	58201	223539
914095	19744000	359787	329272	21151	203885
948519	20327000	344532	385097	34131	184759
1057188	18961280	347602	467027	43766	198793

Figure 8.7 – The DataFrame with the count column in a wide format

How it works...

Both approaches used in *step 1* and *step 2* work as expected if no rows are duplicated on the index/key column. In our case, it's the academic_year column. If there were multiple rows for an academic year, let's say the value 2018/19, then you'd need to specify the aggregation_function parameter to specify how you want to aggregate those duplicate rows when using the .pivot() method.

In a similar manner, you need to be careful how you aggregate your data when using the approach seen in *step 2* when there are multiple rows per value in the index/key column. Please refer to the *There's more...* section for an example. Otherwise, it won't matter how you aggregate because there is only one value per academic year. In our example code in *step 2*, I could have used other aggregation methods such as min and max and they would've worked as well.

The .unstack() method alone doesn't provide the same result as other approaches, but you can imagine how you'd achieve it by using it multiple times on other columns and combining them. The .unstack() method is likely faster than any other approaches, such as .pivot(), .group_by(), .agg(), because .unstack() skips the grouping phase.

One thing worth noting is that the .unstack() method is considered *unstable* as of the time of writing. This means that the functionality may change without it being considered a breaking change, as noted in the Polars API doc: https://docs.pola.rs/py-polars/html/reference/dataframe/api/polars.DataFrame.unstack.html.

> **Tip**
> Both the .pivot() and .unstack() methods are only available as DataFrame methods. If you want to apply the operation on a LazyFrame, you'd need to either use the approach introduced in *step 2* in the recipe or convert your LazyFrame to a DataFrame to apply the .pivot() method. You can convert it back to a LazyFrame with the .lazy() method if necessary.

There's more...

If data was duplicated on the index/key column, here's how you would use `.pivot()` and `.group_by() + .agg()` instead.

Let's prepare such a DataFrame using the `df` variable created at the beginning of this chapter instead of `long_df`:

```
wide_df_with_dups = (
    pl.concat([
        df.head(1),
        df
    ])
)
wide_df_with_dups
```

shape: (6, 7)

academic_year	students	us_students	undergraduate	graduate	non_degree	opt
str	i64	i64	i64	i64	i64	i64
"2018/19"	1095299	19828000	431930	377943	62341	223085
"2018/19"	1095299	19828000	431930	377943	62341	223085
"2019/20"	1075496	19720000	419321	374435	58201	223539
"2020/21"	914095	19744000	359787	329272	21151	203885
"2021/22"	948519	20327000	344532	385097	34131	184759
"2022/23"	1057188	18961280	347602	467027	43766	198793

Figure 8.8 – The DataFrame with duplicated rows

We'll turn this DataFrame into a long format and then see how we can convert it back into a wide format. When you try to use `.pivot()` in the same way as *step 1*, as in the following code, you'll get an error message telling you to specify an aggregation function.

Here is how to convert a wide format into a long format:

```
long_df_with_dups = (
    wide_df_with_dups
    .unpivot(
        index='academic_year',
        on=[
        'students',
        'us_students',
```

```
        'undergraduate',
        'graduate',
        'non_degree',
        'opt'
    ],
    variable_name='student_type',
    value_name='count'
    )
)
```

Apply the pivot operation to convert it back to a wide format:

```
(
    long_df_with_dups
    .pivot(
        index='academic_year',
        values='count',
        on='student_type'
    )
)
```

The preceding code will return the following error:

```
ComputeError: found multiple elements in the same group, please
specify an aggregation function
```

Let's add the `aggregation_function` method to resolve the issue:

```
(
    long_df_with_dups
    .pivot(
        index='academic_year',
        values='count',
        on='student_type',
        aggregate_function='min'
    )
)
```

The preceding code will return the following output:

shape: (5, 7)

academic_year	students	us_students	undergraduate	graduate	non_degree	opt
str	i64	i64	i64	i64	i64	i64
"2018/19"	1095299	19828000	431930	377943	62341	223085
"2019/20"	1075496	19720000	419321	374435	58201	223539
"2020/21"	914095	19744000	359787	329272	21151	203885
"2021/22"	948519	20327000	344532	385097	34131	184759
"2022/23"	1057188	18961280	347602	467027	43766	198793

Figure 8.9 – The DataFrame with duplicate rows removed

By using the `aggregate_function` parameter, Polars groups the DataFrame by the `index` column, which removes the duplicate row from the DataFrame.

Now, if those duplicate rows had different values, then you'd need to carefully determine what aggregation should be used in your specific case.

If none of the aggregation methods will work, but you still need to convert a long format into a wide format, you can still do that using the `list` data type. Let me show you what I mean with the following code:

```
(
    long_df_with_dups
    .pivot(
        index='academic_year',
        values='count',
        on='student_type',
        aggregate_function=pl.element()
    )
)
```

The preceding code will return the following output:

shape: (5, 7)

academic_year	students	us_students	undergraduate	graduate	non_degree	opt
str	list[i64]	list[i64]	list[i64]	list[i64]	list[i64]	list[i64]
"2018/19"	[1095299, 1095299]	[19828000, 19828000]	[431930, 431930]	[377943, 377943]	[62341, 62341]	[223085, 223085]
"2019/20"	[1075496]	[19720000]	[419321]	[374435]	[58201]	[223539]
"2020/21"	[914095]	[19744000]	[359787]	[329272]	[21151]	[203885]
"2021/22"	[948519]	[20327000]	[344532]	[385097]	[34131]	[184759]
"2022/23"	[1057188]	[18961280]	[347602]	[467027]	[43766]	[198793]

Figure 8.10 – The DataFrame with values in lists

Notice that you can inject an expression in the `aggregate_function` parameter instead of an aggregate function.

You can do the same by using the `.group_by()` and `.agg()` methods:

```
agg_cols = [
    (
        pl.col('count')
        .filter(pl.col('student_type')==stu_type)
        .alias(stu_type)
    )
    for stu_type in student_types
]

(
    long_df_with_dups
    .group_by('academic_year', maintain_order=True)
    .agg(agg_cols)
)
```

The preceding code will return the same output as *Figure 8.10*. The columns may be different.

From here, you can apply list operations and other additional transformations as per your needs.

See also

Please refer to these resources to learn more about turning rows into columns:

* https://docs.pola.rs/api/python/stable/reference/dataframe/api/polars.DataFrame.pivot.html
* https://docs.pola.rs/api/python/stable/reference/dataframe/api/polars.DataFrame.unstack.html

Joining DataFrames

A join operation is used to merge rows from two or more datasets by utilizing a shared column that establishes a relationship between them. You may already be familiar with the use and concept of joining, but it's commonly used in any data processing tools such as SQL and other DataFrame libraries such as pandas and Spark.

In this recipe, we'll look at how to apply join operations in Polars DataFrames.

Getting ready

We'll continuously use the same data we've used in previous recipes in this chapter. Execute the following code to do the same process and rename the DataFrame accordingly:

```
from polars import selectors as cs

academic_df = (
```

```
    pl.read_csv('../data/academic.csv')
    .select(
        pl.col('year').alias('academic_year'),
        cs.numeric().cast(pl.Int64)
    )
    .filter(pl.col('academic_year').str.slice(0,4).cast(pl.
Int32)>=2018)
)
```

Also, we'll use another set of data to join onto the other DataFrame. Run the following code to read the data into a DataFrame:

```
status_df = (
    pl.read_csv('../data/status.csv')
    .with_columns(
        cs.float().cast(pl.Int64)
    )
)
status_df
```

The DataFrame contains the following data:

shape: (16, 10)

year	female	male	single	married	full_time	part_time	visa_f	visa_j	visa_other
str	i64	i64	i64	i64	i64	i64	i64	i64	i64
"2007/08"	278841	344964	543958	79847	575772	48033	552691	31814	39300
"2008/09"	304242	367374	591694	79922	613185	58431	589007	39625	42984
"2009/10"	309534	381389	615612	75311	637722	53201	612158	38692	40073
"2010/11"	322582	400695	653842	69435	669031	54246	645163	40504	37610
"2011/12"	338671	425824	690339	74156	714038	50457	688810	42047	33638
"2012/13"	363922	455722	744237	75407	769646	49998	747515	42621	29508
"2013/14"	390749	495303	806307	79745	829345	56707	804535	49619	31898
"2014/15"	426043	548883	886208	88718	912531	62395	881333	58496	35097
"2015/16"	451982	591857	950937	92902	971814	72025	957200	52192	34447
"2016/17"	470366	608456	970940	107882	1007620	71202	994674	45311	38837
"2017/18"	477329	617463	999545	95247	1028010	66782	1015967	44886	33939
"2018/19"	480836	614463	1002199	93100	1030676	64623	1018628	44907	31764
"2019/20"	477520	597976	987305	88191	1016344	59152	1000211	44095	31190
"2020/21"	407686	506409	827256	86839	843710	70385	860163	16454	37478
"2021/22"	428731	519788	861255	87264	884020	64499	884020	27507	36992
"2022/23"	473620	583568	966270	90918	986356	70832	989528	34887	32773

Figure 8.11 – The status_df DataFrame

This `status.csv` file contains student counts broken down by attributes such as gender, marital status, and visa status. Notice it's in a wide format rather than a long format.

How to do it...

Here's how to join data:

1. Use the `.join()` method to combine two DataFrames. We'll join `status_df` onto `academic_df`, leaving only columns relating to visa types:

```
joined_df = (
    academic_df
    .join(
        status_df,
        left_on='academic_year',
        right_on='year',
        how='inner'
    )
    .select(
        'academic_year',
        'students',
        cs.contains('visa')
    )
)
joined_df.head()
```

The preceding code will return the following output:

shape: (5, 5)

academic_year	students	visa_f	visa_j	visa_other
str	i64	i64	i64	i64
"2018/19"	1095299	1018628	44907	31764
"2019/20"	1075496	1000211	44095	31190
"2020/21"	914095	860163	16454	37478
"2021/22"	948519	884020	27507	36992
"2022/23"	1057188	989528	34887	32773

Figure 8.12 – The joined DataFrame

2. Now that we have brought in student counts for each visa type, we can visualize the data and find some trends. To analyze the data by visa types, we'll need to convert the DataFrame into a long format first so that we can treat all visa types altogether in one column, which can then be used to slice the data. We'll look at the percent of total for each visa type:

```python
import plotly.express as px

viz_df = (
    joined_df
    .unpivot(
        index=['academic_year', 'students'],
        on=cs.contains('visa'),
        variable_name='visa_type',
        value_name='count'
    )
    .with_columns(
        (pl.col('count') / pl.col('students')).alias('percent_
of_total')
    )
)

fig = px.bar(
    x=viz_df['academic_year'],
    y=viz_df['percent_of_total'],
    color=viz_df['visa_type'],
    barmode = 'stack',
    text_auto='.1%',
    title='International Student Count by Visa Type',
    labels={
        'x': 'Year',
        'y': 'Student Count'
    },
)

fig.update_layout(
    autosize=True,
    uniformtext_minsize=10,
    uniformtext_mode='hide',
    yaxis_tickformat = '0%'
)

fig.update_traces(textposition='inside')

fig.show()
```

The preceding code will return the following output:

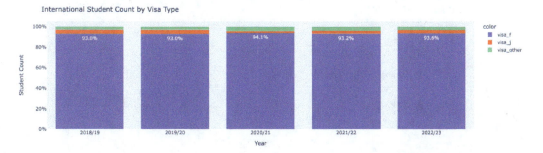

Figure 8.13 – International student counts percent of total by visa types

3. Utilize the `validate` parameter to ensure the cardinality of columns you join on.

    Let's say `status_df` comes in a long format. The following DataFrame will have the level of grain where each row represents data per year per visa type. There will be multiple rows per year value:

```
status_long_df = (
    status_df
    .unpivot(
        index='year',
        on=cs.contains('visa'),
        variable_name='visa_type',
        value_name='count'
    )
)
status_long_df.head()
```

The preceding code will produce the following output:

shape: (5, 3)

year	visa_type	count
str	str	i64
"2007/08"	"visa_f"	552691
"2008/09"	"visa_f"	589007
"2009/10"	"visa_f"	612158
"2010/11"	"visa_f"	645163
"2011/12"	"visa_f"	688810

Figure 8.14 – status_df in a long format

4. Now, let's join two DataFrames with the `validate` parameter:

```
(
    academic_df
    .join(
        status_long_df,
        left_on='academic_year',
        right_on='year',
        how='inner',
        validate='1:m'
    )
    .select(
        'academic_year',
        'students',
        'visa_type',
        'count'
    )
)
```

The preceding code will return the following output:

shape: (15, 4)

academic_year	students	visa_type	count
str	i64	str	i64
"2018/19"	1095299	"visa_f"	1018628
"2019/20"	1075496	"visa_f"	1000211
"2020/21"	914095	"visa_f"	860163
"2021/22"	948519	"visa_f"	884020
"2022/23"	1057188	"visa_f"	989528
"2018/19"	1095299	"visa_j"	44907
"2019/20"	1075496	"visa_j"	44095
"2020/21"	914095	"visa_j"	16454
"2021/22"	948519	"visa_j"	27507
"2022/23"	1057188	"visa_j"	34887
"2018/19"	1095299	"visa_other"	31764
"2019/20"	1075496	"visa_other"	31190
"2020/21"	914095	"visa_other"	37478
"2021/22"	948519	"visa_other"	36992
"2022/23"	1057188	"visa_other"	32773

Figure 8.15 – The joined DataFrame with status_df in a long format

The following code will raise an error since the join type is specified as 1 : 1, resulting in a validation failure:

```
(
    academic_df
    .join(
        status_long_df,
        left_on='academic_year',
        right_on='year',
        how='inner',
        validate='1:1'
    )
    .select(
        'academic_year',
        'students',
        'visa_type',
        'count'
    )
)
```

The preceding code will return the following output:

```
>> ComputeError: the join keys did not fulfil 1:1 validation
```

How it works...

The `.join()` method provides parameters to specify columns and additional configurations for your join. I used the `left_on` and `right_on` parameters to specify join columns as they had different names across two DataFrames. I could've used the `on` parameter instead if they had the same name, which is usually the case if your data is cleaned and validated in upstream processes.

There are several join types available, such as `inner`, `left`, `outer`, `semi`, `anti`, `cross`, and `outer_coalesce`. To learn more about join types, please refer to the resources in the *See also* section.

The `validate` parameter helps check the uniqueness of the join keys/columns. As shown in *step 2*, it'll raise an error if the validation fails.

> **Important**
>
> As of the time of writing, the `validate` parameter is not supported in the streaming engine. Also, it only works when joined with single columns.

You can apply the join operation in LazyFrames as well as in DataFrames.

> **Note**
>
> The `coalesce` parameter for a left join decides whether to keep the join column(s) from the right table or not. Only when you set it to False, you'll get the join column with the _right suffix.

There's more...

You can apply an `asof` join in Polars. It is like a left join, but instead of matching on equal keys, it matches based on the closest key. Here's a simple example of how an `asof` join works:

```
a = pl.DataFrame({
    'int': [1,2,3], 'value': [10,20,30]
}).set_sorted('int')

b = pl.DataFrame({
    'int': [4,5,6]
}).set_sorted('int')

b.join_asof(a, on='int', strategy='backward')
```

The preceding code will return the following output:

shape: (3, 2)

int	value
i64	i64
4	30
5	30
6	30

Figure 8.16 – The joined DataFrame using an asof join

Note that you need to set the sorting for each DataFrame for this to work. For more information, please refer to the Polars documentation page linked in the *See also* section.

See also

To learn more about joins in Polars, please use the following resources:

- https://docs.pola.rs/api/python/stable/reference/dataframe/api/polars.DataFrame.join.html

- https://docs.pola.rs/api/python/stable/reference/dataframe/api/polars.DataFrame.join_asof.html

- https://docs.pola.rs/user-guide/transformations/joins/

Concatenating DataFrames

Concatenation in data means combining multiple assets, whether it be a table or DataFrame, into one. We also use it often as in concatenating strings. In the context of concatenating DataFrames, it means combining multiple DataFrames across rows or columns. Unlike the join operation, concatenation doesn't require key columns.

In this recipe, we'll look at how we can concatenate DataFrames in Python Polars.

Getting ready

We'll be using the `academic_df` DataFrame that we already created in the last recipe. Let's split the DataFrame into a few DataFrames so that we can test the concat operation:

```
df1 = academic_df.head(2)
df2 = academic_df.slice(2, 2)
df3 = academic_df.tail(1)
display(df1, df2, df3)
```

The preceding code will return the following output:

shape: (2, 7)

academic_year	students	us_students	undergraduate	graduate	non_degree	opt
str	i64	i64	i64	i64	i64	i64
"2018/19"	1095299	19828000	431930	377943	62341	223085
"2019/20"	1075496	19720000	419321	374435	58201	223539

shape: (2, 7)

academic_year	students	us_students	undergraduate	graduate	non_degree	opt
str	i64	i64	i64	i64	i64	i64
"2020/21"	914095	19744000	359787	329272	21151	203885
"2021/22"	948519	20327000	344532	385097	34131	184759

shape: (1, 7)

academic_year	students	us_students	undergraduate	graduate	non_degree	opt
str	i64	i64	i64	i64	i64	i64
"2022/23"	1057188	18961280	347602	467027	43766	198793

Figure 8.17 – The DataFrame split into two DataFrames

All of the DataFrames have the same number of columns. Only `df1` and `df2` have two rows, and `df3` has one row.

Note that how we split the DataFrame doesn't have a purpose in analyzing the data. It's for demonstration purposes only.

How to do it...

Here's how we can concatenate DataFrames:

1. Concatenate the DataFrames vertically, or union the DataFrames using the `pl.concat()` function:

```
pl.concat(
    [df1, df2, df3],
    how='vertical'
)
```

The preceding code will return the following output:

shape: (5, 7)

academic_year	students	us_students	undergraduate	graduate	non_degree	opt
str	i64	i64	i64	i64	i64	i64
"2018/19"	1095299	19828000	431930	377943	62341	223085
"2019/20"	1075496	19720000	419321	374435	58201	223539
"2020/21"	914095	19744000	359787	329272	21151	203885
"2021/22"	948519	20327000	344532	385097	34131	184759
"2022/23"	1057188	18961280	347602	467027	43766	198793

Figure 8.18 – The vertically concatenated DataFrame

2. Now, repeat what we performed under *step 1* using the `.vstack()` method:

```
df1.vstack(df2).vstack(df3)
```

The preceding code will return the same output as *Figure 8.18*.

3. Concatenate the DataFrames horizontally with `pl.concat()`:

```
pl.concat(
    [
        df1.select('academic_year'),
        df2.select('students')
    ],
    how='horizontal'
)
```

The preceding code will return the following output:

shape: (2, 2)

academic_year	students
str	i64
"2018/19"	914095
"2019/20"	948519

Figure 8.19 – The horizontally concatenated DataFrame

You can stack a DataFrame that has a different length than other DataFrames participating in the concatenation. If you remember, df1 and df2 have two rows, but df3 has only one row. Let's see how Polars inserts null values to fill missing rows automatically:

```
pl.concat(
    [
        df1.select('academic_year'),
        df2.select('students'),
        df3.select('us_students')
    ],
    how='horizontal'
)
```

The preceding code will return the following output:

shape: (2, 3)

academic_year	students	us_students
str	i64	i64
"2018/19"	914095	18961280
"2019/20"	948519	null

Figure 8.20 – The concatenated DataFrame with a missing row supplemented with null

4. Now, let's repeat what we performed under *step 3* using .hstack():

```
(
    df1
    .select('academic_year')
    .hstack(df2.select('students'))
)
```

The preceding code will return the same output as *Figure 8.19*.

Unlike the `pl.concat()` function, you cannot have a DataFrame with a different length with the `.hstack()` method. The following code will result in an error:

```
(
    df1
    .select('academic_year')
    .hstack(df2.select('students'))
    .hstack(df3.select('us_students'))
)
```

The preceding code will produce the following error:

```
>> ShapeError: could not create a new DataFrame: series
"academic_year" has length 2 while series "us_students" has
length 1
```

How it works...

Both the `pl.concat()` function and the `.vstack()` and `.hstack()` methods serve a similar purpose of concatenating DataFrames, but they're different in that `pl.concat()` accepts DataFrames, LazyFrames, and even Series objects as input, whereas `.vstack()` and `.hstack()` only work on DataFrames.

> **Tip**
> Both the `.vstack()` and `.hstack()` methods have the `in_place` parameter as of the time of writing. But I don't recommend using it as it makes your code confusing and harder to debug and hinders method chaining.

There's more...

The `pl.concat()` function has a few other variations of concatenations, including `diagonal`, `vertical_relaxed`, `diagonal_relaxed`, and align. To learn about these options, please visit the Polars documentation page: `https://docs.pola.rs/api/python/stable/reference/api/polars.concat.html`.

The `.extend()` DataFrame method can be used in the same way as you use the `.vstack()` method:

```
df1.extend(df2).extend(df3)
```

The preceding code will return the following output:

shape: (5, 7)

academic_year	students	us_students	undergraduate	graduate	non_degree	opt
str	i64	i64	i64	i64	i64	i64
"2018/19"	1095299	19828000	431930	377943	62341	223085
"2019/20"	1075496	19720000	419321	374435	58201	223539
"2020/21"	914095	19744000	359787	329272	21151	203885
"2021/22"	948519	20327000	344532	385097	34131	184759
"2022/23"	1057188	18961280	347602	467027	43766	198793

Figure 8.21 – The vertically extended DataFrame

Note that just like the `.vstack()` method, this method is not available for LazyFrames. Also, this method works "in place," meaning it changes the DataFrame on the fly without assigning the result to a variable. If you look at `df1` now, it returns the same output as *Figure 8.21*. So, you need to be cautious when using the `.extend()` method.

See also

Please use the following resources to learn more about concatenating and stacking DataFrames:

- `https://docs.pola.rs/api/python/stable/reference/api/polars.concat.html`

- `https://docs.pola.rs/api/python/stable/reference/dataframe/api/polars.DataFrame.vstack.html`

- `https://docs.pola.rs/api/python/stable/reference/dataframe/api/polars.DataFrame.hstack.html`

- `https://docs.pola.rs/api/python/stable/reference/dataframe/api/polars.DataFrame.extend.html`

- `https://docs.pola.rs/user-guide/transformations/concatenation/`

Other techniques for reshaping data

In this chapter, we have so far covered crucial data operations such as pivoting, unpivoting/unpivoting, joining, and concatenating. There are a few other techniques you can use to reshape your data, such as `.partition_by()`, `.transpose()`, `.reshape()`. We'll cover those methods in this recipe.

Getting ready

We'll continue to use the `academic_df` DataFrame that we've become familiar with throughout this chapter.

How to do it...

Here's how to apply those reshaping techniques:

1. To partition the DataFrame into separate DataFrames by column values, we used the `.partition_by()` method:

```
academic_df.partition_by('academic_year')
```

The preceding code will return the following output:

[shape: (1, 7)

academic_year	students	us_students	undergraduate	graduate	non_degree	opt
str	i64	i64	i64	i64	i64	i64
2018/19	1095299	19828000	431930	377943	62341	223085

shape: (1, 7)

academic_year	students	us_students	undergraduate	graduate	non_degree	opt
str	i64	i64	i64	i64	i64	i64
2019/20	1075496	19720000	419321	374435	58201	223539

shape: (1, 7)

academic_year	students	us_students	undergraduate	graduate	non_degree	opt
str	i64	i64	i64	i64	i64	i64
2020/21	914095	19744000	359787	329272	21151	203885

shape: (1, 7)

academic_year	students	us_students	undergraduate	graduate	non_degree	opt
str	i64	i64	i64	i64	i64	i64
2021/22	948519	20327000	344532	385097	34131	184759

shape: (1, 7)

academic_year	students	us_students	undergraduate	graduate	non_degree	opt
str	i64	i64	i64	i64	i64	i64
2022/23	1057188	18961280	347602	467027	43766	198793

Figure 8.22 – The partitioned DataFrames by values in the academic_year column

2. Use the `.transpose()` method to flip rows and columns:

```
academic_df.transpose(include_header=True)
```

The preceding code will return the following output:

shape: (7, 6)

column	column_0	column_1	column_2	column_3	column_4
str	str	str	str	str	str
"academic_year"	"2018/19"	"2019/20"	"2020/21"	"2021/22"	"2022/23"
"students"	"1095299"	"1075496"	"914095"	"948519"	"1057188"
"us_students"	"19828000"	"19720000"	"19744000"	"20327000"	"18961280"
"undergraduate"	"431930"	"419321"	"359787"	"344532"	"347602"
"graduate"	"377943"	"374435"	"329272"	"385097"	"467027"
"non_degree"	"62341"	"58201"	"21151"	"34131"	"43766"
"opt"	"223085"	"223539"	"203885"	"184759"	"198793"

Figure 8.23 – The transposed DataFrame

3. Reshape column values into a list using the `.reshape()` method:

```
(
    academic_df
    .select(
        pl.col('academic_year', 'students').reshape((1, 5))
    )
)
```

The preceding code will return the following output:

shape: (1, 2)

academic_year	students
array[str, 5]	array[i64, 5]
["2018/19", "2019/20", ... "2022/23"]	[1095299, 1075496, ... 1057188]

Figure 8.24 – Column values reshaped into a list

How it works...

The `.partition_by()` method can also return the result as dictionaries as well as DataFrames. Also, you can specify multiple columns by which you group the data.

The `.transpose()` method helps flip the rows and columns in a DataFrame. You can specify the header name and column names with the `header_name` and `column_names` parameters.

The `.reshape()` method turns column values into an array. You specify the shape that you want as a tuple as a parameter in the method. The first value in the tuple would be the number of rows, and the second value you specify becomes the number of elements in the list. In the preceding example, it was specified that the resulting DataFrame would have one row with five array elements.

> **Note**
>
> As noted in the preceding chapter, an array is a data type in which each row represents a fixed-size array of elements with the same data type. A list is a data type in which each row represents a list of elements with the same data type but can have different lengths.

Finally, both the `.partition_by()` and `.transpose()` methods work only on DataFrames, and the `.reshape()` method only works on a column or a Series object.

See also

To learn more about the techniques introduced in this recipe, please refer to the following resources:

- https://docs.pola.rs/api/python/stable/reference/dataframe/api/polars.DataFrame.partition_by.html

- https://docs.pola.rs/api/python/stable/reference/dataframe/api/polars.DataFrame.transpose.html

- https://docs.pola.rs/api/python/stable/reference/expressions/api/polars.Expr.reshape.html

9

Time Series Analysis

Time series analysis is one of the most common and impactful things you can do with your time series data. So far in this book, we've covered many techniques in Python Polars in regards to transforming, manipulating, and analyzing data. In this chapter, you will continue to learn about what Python Polars is capable of. This chapter teaches you how to work with date and time columns. You will also learn to identify trends and seasonality in your data using various methods. You will build calculations on time series data, including a time series forecasting model to predict future values. Time series analysis provides you and your organization with insights and meaningful statistics from data. This chapter delves into how we can work with time series data and conduct analysis on it, leveraging the capability of Python Polars.

In this chapter, we're going to cover the following main topics:

- Working with date and time
- Applying rolling windows calculations
- Resampling techniques
- Time series forecasting with the functime library

Technical requirements

You can download the datasets and code from the GitHub repository as follows:

- Data: `https://github.com/PacktPublishing/Polars-Cookbook/tree/main/data`
- Code: `https://github.com/PacktPublishing/Polars-Cookbook/tree/main/Chapter09`

It is assumed that you have installed the Polars library in your Python environment:

```
>>> pip install polars
```

And that you have imported it into your code:

```
import polars as pl
```

We'll also be using the Polars built-in visualization feature in this chapter. Make sure to import the hvplot library in the command line:

```
>>> pip install hvplot
```

We'll be using the historical hourly weather dataset throughout this chapter. The dataset contains weather data in Toronto such as temperature and humidity. You can find it at this URL in our GitHub repo: https://github.com/PacktPublishing/Polars-Cookbook/blob/main/data/toronto_weather.csv.

We'll use a LazyFrame instead of a DataFrame throughout this chapter as you'll probably use a LazyFrame for most of your projects at work to enable more efficient data processing. Read the dataset into a LazyFrame with the following code:

```
lf = pl.scan_csv('../data/toronto_weather.csv')
```

Take a look at the first five rows:

```
lf.head().collect()
```

The preceding code will return the following output:

shape: (5, 5)

datetime	temperature	wind_speed	pressure	humidity
str	f64	f64	f64	f64
"2012-10-01 12:..."	null	null	null	null
"2012-10-01 13:..."	286.26	3.0	1012.0	82.0
"2012-10-01 14:..."	286.262541	3.0	1011.0	81.0
"2012-10-01 15:..."	286.269518	3.0	1011.0	79.0
"2012-10-01 16:..."	286.276496	3.0	1010.0	77.0

Figure 9.1 – The first five rows in the Toronto weather dataset

Since the temperature is in Kelvin, let's convert it to Celsius (sorry for those who use Fahrenheit, but most of the countries on Earth use Celsius):

```
lf = lf.with_columns(pl.col('temperature')-273.15)
```

Now we made the change, let's look at the DataFrame again:

```
lf.head().collect()
```

The preceding code will return the following output:

shape: (5, 5)

datetime	temperature	wind_speed	pressure	humidity
str	f64	f64	f64	f64
"2012-10-01 12:..."	null	null	null	null
"2012-10-01 13:..."	13.11	3.0	1012.0	82.0
"2012-10-01 14:..."	13.112541	3.0	1011.0	81.0
"2012-10-01 15:..."	13.119518	3.0	1011.0	79.0
"2012-10-01 16:..."	13.126496	3.0	1010.0	77.0

Figure 9.2 – The temperature column converted to Celsius

Regarding the `.fetch()` method that was available until Polars version 0.20.31, I shared the following tip in *Chapter 1*, *Getting Started with Python Polars*, however, as I think it's worth repeating the implications of the method, I decided to include it in this chapter as well.

> **Important**
>
> You could use the `._fetch()` method for a LazyFrame just as you use the `.head()` method for a DataFrame. The `._fetch()` method used to be `.fetch()`, however, since Polars version 1.0.0, the method became only for Polars project development, and you can no longer find the page about the method in Polars API documentation. For whatever reason, if you'll ever have to use the `._fetch()` method, keep in mind that it doesn't guarantee the final number of rows. Let's say you scan the specified number of rows in a LazyFrame and you have subsequent operations such as filtering, aggregations, and joins. Then those operations are executed only on those scanned rows. If you specify a small number of rows in `._fetch()`, you can see that you'll most likely get the wrong output.
>
> As Polars library users, we'll only need to use the combination of `.head()` and `.collect()`. This ensures that Polars materializes data as you apply your subsequent operations and will get you the correct output.

Working with date and time

In time series analysis, you may spend a great deal of time transforming and manipulating date and time columns. This includes parsing dates when upon reading a file, casting strings to date, extracting information from date columns, and utilizing temporal expressions.

Also, it's good to remember that we have time zones across the globe. We'll learn how to work with time zones in this recipe as well.

How to do it...

Here are a few things that may be helpful when working with date and time:

1. If your source data has a date, datetime, or time column of the `string` date type, you may be able to use the `try_parse_date` parameter to convert it to the `date/datetime/time` data type upon reading.

    Let's add the parameter to the `scan_csv()` method:

    ```
    lf_date_parsed = pl.scan_csv('../data/toronto_weather.csv', try_
    parse_dates=True)

    lf_date_parsed.head().collect()
    ```

    The preceding code will return the following output:

    shape: (5, 5)

datetime	temperature	wind_speed	pressure	humidity
datetime[µs]	f64	f64	f64	f64
2012-10-01 12:00:00	null	null	null	null
2012-10-01 13:00:00	286.26	3.0	1012.0	82.0
2012-10-01 14:00:00	286.262541	3.0	1011.0	81.0
2012-10-01 15:00:00	286.269518	3.0	1011.0	79.0
2012-10-01 16:00:00	286.276496	3.0	1010.0	77.0

    Figure 9.3 – The LazyFrame with the string column converted to datetime

    Notice that the data type of the `datetime` column is now in datetime.

    You can also check the data types of columns using the `.schema` or `.dtypes` attributes:

    ```
    lf_date_parsed.collect_schema(),
    lf_date_parsed.collect_schema().dtypes()
    ```

    The preceding code will return the following output:

    ```
    (Schema([('datetime', Datetime(time_unit='us', time_zone=None)),
            ('temperature', Float64),
            ('wind_speed', Float64),
    ```

```
        ('pressure', Float64),
        ('humidity', Float64)]),
  [Datetime(time_unit='us', time_zone=None),
  Float64,
  Float64,
  Float64,
  Float64])
```

2. You can convert the data type after reading the data into a DataFrame as well:

```
lf = lf.with_columns(pl.col('datetime').str.to_datetime())
lf.head().collect()
```

The preceding code will return the same output as *Figure 9.3*.

3. Now, we're ready to extract date features such as the time, day, week, month, and year from the `datetime` column:

```
(
    lf
    .select(
        'datetime',
        pl.col('datetime').dt.year().alias('year'),
        pl.col('datetime').dt.month().alias('month'),
        pl.col('datetime').dt.day().alias('day'),
        pl.col('datetime').dt.time().alias('time')
    )
    .head().collect()
)
```

The preceding code will return the following output:

shape: (5, 5)

datetime	year	month	day	time
datetime[µs]	i32	i8	i8	time
2012-10-01 12:00:00	2012	10	1	12:00:00
2012-10-01 13:00:00	2012	10	1	13:00:00
2012-10-01 14:00:00	2012	10	1	14:00:00
2012-10-01 15:00:00	2012	10	1	15:00:00
2012-10-01 16:00:00	2012	10	1	16:00:00

Figure 9.4 – Date and time attributes added to the LazyFrame

4. Filter on date and time attributes:

```
from datetime import datetime

filtered_lf = (
    lf
    .filter(
        pl.col('datetime').dt.date().is_between(
            datetime(2017,1,1), datetime(2017,12,31)
        ),
        pl.col('datetime').dt.hour() < 12
    )
)
filtered_lf.head().collect()
```

The preceding code will return the following output:

shape: (5, 5)

datetime	temperature	wind_speed	pressure	humidity
datetime[µs]	f64	f64	f64	f64
2017-01-01 00:00:00	2.44	3.0	1001.0	92.0
2017-01-01 01:00:00	2.19	9.0	1003.0	86.0
2017-01-01 02:00:00	2.41	7.0	1003.0	86.0
2017-01-01 03:00:00	2.42	5.0	1003.0	74.0
2017-01-01 04:00:00	1.77	6.0	1006.0	69.0

Figure 9.5 – The output LazyFrame with data filtered on a datetime column

Let's check that the data is filtered correctly by looking at the unique values in the year and hour attributes of the datetime column. We're adding extra logic to the preceding code:

```
(
    filtered_lf
    .select(
        pl.col('datetime').dt.year().unique()
        .implode()
        .list.len()
        .alias('year_cnt'),
        pl.col('datetime').dt.hour().unique()
        .implode()
        .list.len()
```

```
                .alias('hour_cnt')
        )
        .head()
        .collect()
    )
```

The preceding code will return the following output:

shape: (1, 2)

year_cnt	hour_cnt
u32	u32
1	12

Figure 9.6 – The unique count of year and hour in the datetime column after filtering

5. Convert and replace the time zones in your `date` column:

```
time_zones_lf = (
    lf
    .select(
        'datetime',
        pl.col('datetime').dt.replace_time_zone('America/
Toronto')
        .alias('replaced_time_zone_toronto'),
        pl.col('datetime').dt.convert_time_zone('America/
Toronto')
        .alias('converted_time_zone_toronto')
    )
)
time_zones_lf.head().collect()
```

The preceding code will return the following output:

shape: (5, 3)

datetime	replaced_time_zone_toronto	converted_time_zone_toronto
datetime[µs]	datetime[µs, America/Toronto]	datetime[µs, America/Toronto]
2012-10-01 12:00:00	2012-10-01 12:00:00 EDT	2012-10-01 08:00:00 EDT
2012-10-01 13:00:00	2012-10-01 13:00:00 EDT	2012-10-01 09:00:00 EDT
2012-10-01 14:00:00	2012-10-01 14:00:00 EDT	2012-10-01 10:00:00 EDT
2012-10-01 15:00:00	2012-10-01 15:00:00 EDT	2012-10-01 11:00:00 EDT
2012-10-01 16:00:00	2012-10-01 16:00:00 EDT	2012-10-01 12:00:00 EDT

Figure 9.7 – The LazyFrame with columns with the time zone replaced or converted

How it works...

The `dt` namespace in the Expression API provides a variety of methods to work with date and time attributes.

Being able to both convert upon reading the file and convert data types after reading data in a DataFrame or LazyFrame gives you a lot of flexibility. For example, in *step 2*, we could've used the `.str.strptime()` method instead of the `.str.to_datetime()` method to convert a string column to a `datetime` column.

Also, on top of the comparison operators such as "greater than", "less than", and so on, the `.is_between()` method can be handy for specifying a date range. Python Datetime objects can be used to filter date and time attributes.

In *step 4*, we used the `.implode()` method to aggregate rows into lists. It's useful to reshape your DataFrame or LazyFrame for transformations.

The two main methods to work with time zones are the `.dt.replace_time_zone()` and `.dt.convert_time_zone()`. The difference between them is that the former doesn't convert the datetime values; it just assigns or resets the time zone of the column, whereas the latter modifies the datetime values while changing the time zone of the column.

> **Important**
>
> You should not mess with time zones unless you absolutely have to. In other words, make sure you understand all the details of the dataset and the date column you're trying to convert the time zone for.

There is more...

One useful function you can use to offset your date and time attributes is `pl.duration()`. Look at the following example to see how you can use it in adjusting your data:

```
(
    lf
    .select(
        'datetime',
        (pl.col('datetime')-pl.duration(weeks=5)).
alias('minus_5weeks'),

        (pl.col('datetime')+pl.duration(milliseconds=5)).
alias('plus_5ms'),

    )
    .head()
    .collect()
)
```

The preceding code will return the following output:

shape: (5, 3)

datetime	minus_5weeks	plus_5ms
datetime[µs]	datetime[µs]	datetime[µs]
2012-10-01 12:00:00	2012-08-27 12:00:00	2012-10-01 12:00:00.005
2012-10-01 13:00:00	2012-08-27 13:00:00	2012-10-01 13:00:00.005
2012-10-01 14:00:00	2012-08-27 14:00:00	2012-10-01 14:00:00.005
2012-10-01 15:00:00	2012-08-27 15:00:00	2012-10-01 15:00:00.005
2012-10-01 16:00:00	2012-08-27 16:00:00	2012-10-01 16:00:00.005

Figure 9.8 – Offsetting datetime values with pl.duration

See also

Here are a few helpful Polars documentation pages:

- `https://docs.pola.rs/user-guide/transformations/time-series/parsing/`

- `https://docs.pola.rs/user-guide/transformations/time-series/timezones/`

- `https://docs.pola.rs/py-polars/html/reference/expressions/temporal.html`

- `https://docs.pola.rs/py-polars/html/reference/expressions/api/polars.duration.html`

Applying rolling window calculations

Rolling calculations such as rolling sum and average are essential for grasping the dynamics within time series data, offering valuable insights into trends, patterns, and anomalies across varying time spans. Rolling calculations serve as a fundamental tool in the analysis of time series data.

In this recipe, we'll look at how to apply rolling calculations using Polars' built-in methods.

How to do it...

Here's how to apply rolling calculations.

1. Let's see how you can calculate the rolling average of the temperature using the built-in `.rolling_mean()` method:

```
(
    lf
    .select(
        'datetime',
        'temperature',
        pl.col('temperature').rolling_mean(3).alias('3hr_
rolling_avg')
    )
    .head()
    .collect()
)
```

The preceding code will return the following output:

shape: (5, 3)

datetime	temperature	3hr_rollign_avg
datetime[µs]	f64	f64
2012-10-01 12:00:00	null	null
2012-10-01 13:00:00	13.11	null
2012-10-01 14:00:00	13.112541	null
2012-10-01 15:00:00	13.119518	13.11402
2012-10-01 16:00:00	13.126496	13.119518

Figure 9.9 – Rolling average calculation in a 3-hr window

2. Aggregating the temperature at the hourly level doesn't give you much insight. Let us aggregate the temperature at the day level first and then look at a few rolling calculations:

    First, let's aggregate the data to the day level:

```
daily_avg_temperature_lf = (
    lf
    .select(
        pl.col('datetime').dt.date().alias('date'),
        'temperature'
```

```
    )
    .group_by('date', maintain_order=True)
    .agg(
        pl.col('temperature').mean().alias('daily_avg_temp')
    )
)
daily_avg_temperature_lf.head().collect()
```

The preceding code will return the following output:

shape: (5, 2)

date	daily_avg_temp
date	f64
2012-10-01	13.140854
2012-10-02	14.24739
2012-10-03	14.176875
2012-10-04	15.067917
2012-10-05	16.216458

Figure 9.10 – The LazyFrame aggregated to the day level

3. Next, we'll calculate the rolling aggregations on the daily average temperature:

```
(
    daily_avg_temperature_lf
    .select(
        'date',
        'daily_avg_temp',
        pl.col('daily_avg_temp').rolling_mean(3).alias('3day_
rolling_avg'),

        pl.col('daily_avg_temp').rolling_min(3).alias('3day_
rolling_min'),

        pl.col('daily_avg_temp').rolling_max(3).alias('3day_
rolling_max')

    )
    .head()
    .collect()
)
```

The preceding code will return the following output:

shape: (5, 5)

date	daily_avg_temp	3day_rolling_avg	3day_rolling_min	3day_rolling_max
date	f64	f64	f64	f64
2012-10-01	13.140854	null	null	null
2012-10-02	14.24739	null	null	null
2012-10-03	14.176875	13.85504	13.140854	14.24739
2012-10-04	15.067917	14.497394	14.176875	15.067917
2012-10-05	16.216458	15.15375	14.176875	16.216458

Figure 9.11 – 3-day rolling aggregations are added to the LazyFrame

4. You can utilize the `.rolling()` method instead to give you more flexibility in the aggregation type you want to apply:

```
(
    daily_avg_temperature_lf
    .set_sorted('date')
    .select(
        'date',
        'daily_avg_temp',
        pl.col('daily_avg_temp').rolling_mean(3).alias('3day_
rolling_avg'),
        pl.col('daily_avg_temp').rolling_mean(
            window_size=3,
            min_periods=1
        )
        .alias('3day_rolling_avg2'),
        pl.col('daily_avg_temp').mean().rolling(
            index_column='date',
            period='3d',
            closed='right'
        )
        .alias('3day_rolling_avg3')
    )
    .head(10)
    .collect()
)
```

The preceding code will return the following output:

shape: (10, 5)

date	daily_avg_temp	3day_rolling_avg	3day_rolling_avg2	3day_rolling_avg3
date	f64	f64	f64	f64
2012-10-01	13.140854	null	13.140854	13.140854
2012-10-02	14.24739	null	13.694122	13.694122
2012-10-03	14.176875	13.85504	13.85504	13.85504
2012-10-04	15.067917	14.497394	14.497394	14.497394
2012-10-05	16.216458	15.15375	15.15375	15.15375
2012-10-06	15.725417	15.669931	15.669931	15.669931
2012-10-07	10.197083	14.046319	14.046319	14.046319
2012-10-08	6.79625	10.90625	10.90625	10.90625
2012-10-09	6.735	7.909444	7.909444	7.909444
2012-10-10	9.305417	7.612222	7.612222	7.612222

Figure 9.12 – 3-day rolling averages with two different methods

5. If your DataFrame or LazyFrame only contains rolling calculations, you should utilize the `.rolling()` DataFrame/LazyFrame method as it caches the window size computation:

```
(
    daily_avg_temperature_lf
    .set_sorted('date')
    .rolling(
        'date',
        period='3d'
    )
    .agg(
        pl.col('daily_avg_temp'),
        pl.col('daily_avg_temp').mean().alias('3day_rolling_
avg'),
        pl.col('daily_avg_temp').min().alias('3day_rolling_
min'),
        pl.col('daily_avg_temp').max().alias('3day_rolling_
max'),
    )
    .head(10)
    .collect()
)
```

The preceding code will return the following output:

shape: (10, 5)

date	daily_avg_temp	3day_rolling_avg	3day_rolling_min	3day_rolling_max
date	list[f64]	f64	f64	f64
2012-10-01	[13.140854]	13.140854	13.140854	13.140854
2012-10-02	[13.140854, 14.24739]	13.694122	13.140854	14.24739
2012-10-03	[13.140854, 14.24739, 14.176875]	13.85504	13.140854	14.24739
2012-10-04	[14.24739, 14.176875, 15.067917]	14.497394	14.176875	15.067917
2012-10-05	[14.176875, 15.067917, 16.216458]	15.15375	14.176875	16.216458
2012-10-06	[15.067917, 16.216458, 15.725417]	15.669931	15.067917	16.216458
2012-10-07	[16.216458, 15.725417, 10.197083]	14.046319	10.197083	16.216458
2012-10-08	[15.725417, 10.197083, 6.79625]	10.90625	6.79625	15.725417
2012-10-09	[10.197083, 6.79625, 6.735]	7.909444	6.735	10.197083
2012-10-10	[6.79625, 6.735, 9.305417]	7.612222	6.735	9.305417

Figure 9.13 – 3-day rolling aggregations with the .rolling method at LazyFrame level

6. Now, let's visualize the daily average temperature and the 60-day rolling average. We'll use Polars' built-in plotting functionality:

```
(
    daily_avg_temperature_lf
    .select(
        'date',
        'daily_avg_temp',
        pl.col('daily_avg_temp').rolling_mean(60).alias('60day_
rolling_avg')
    )
    .collect()
    .plot.line(
        x='date',
        y=['daily_avg_temp', '60day_rolling_avg'],
        color=['skyblue', 'gray'],
        width=800,
        height=400
    )
    .opts(legend_position='bottom_right')
)
```

The preceding code will return the following output:

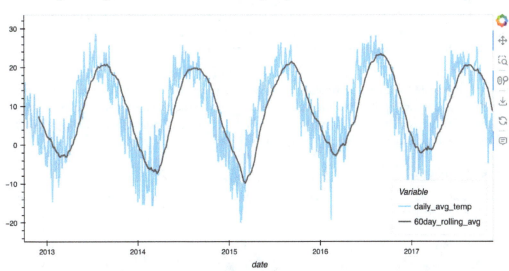

Figure 9.14 – A line plot showing the daily average temperature and 60-day rolling average

As you might've expected, there is definitely a seasonality in the temperatures data.

How it works...

There is nothing magical about calculating rolling aggregations in Polars. There are built-in methods such as `.rolling_mean()`, `.rolling_min()`, and `.rolling_max()`. There is a lot more variety of rolling calculations you can do, including `.rolling_var()` and `.rolling_std()`. You can check out more rolling calculation methods in the Polars documentation page: `https://docs.pola.rs/py-polars/html/reference/expressions/computation.html`.

These methods provide a way to configure the details of your rolling calculations. For example, the `weights` parameter is a list of numbers you can specify that will be used to multiply the output values. The `center` parameter sets the output value in the center of the specified window.

You can also use the `.rolling()` method instead of using the canned expression methods such as `.rolling_mean()`. The rolling method is more flexible in that you can utilize any specified expression in your rolling calculations. As you saw in *step 3*, both `.rolling_mean()` and `.rolling()` accomplish the same thing by adjusting the parameters.

The `.rolling()` DataFrame/LazyFrame method should be utilized if your resulting output only contains rolling calculations. This method caches the window size computation so you can have multiple rolling calculations more efficiently. It makes sense to use other rolling expression methods when you need rolling calculations while keeping some other columns/attributes in your output.

> **Important**
>
> Both of the `.rolling()` methods used at the DataFrame/LazyFrame level, and the expression level require the time column to be sorted. This can be done either with the `.set_sorted()` method if the column was already sorted or the `.sort()` method.

One of the important techniques in time series analysis is to visualize the data based on the date and time-related axis. Polars' built-in plotting functionality comes in handy when you try to understand your data better by visualizing your data. It uses the `hvplot` library under the hood, so you'd need to have it installed in order to utilize its functionality.

There is more...

The `.rolling_map()` method accepts a user-defined function as an argument and apply your custom rolling calculations:

```python
def get_range(nums):
    min_num = min(nums)
    max_num = max(nums)
    range = max(nums) - min(nums)
    return range


(
    daily_avg_temperature_lf
    .with_columns(
        pl.col('daily_avg_temp').rolling_map(get_range, window_size=3).alias('3day_rolling_range')
    )
    .head()
    .collect()
)
```

The preceding code will return the following output:

shape: (5, 3)

date	daily_avg_temp	3day_rolling_range
date	f64	f64
2012-10-01	13.140854	null
2012-10-02	14.24739	null
2012-10-03	14.176875	1.106536
2012-10-04	15.067917	0.891042
2012-10-05	16.216458	2.039583

Figure 9.15 – Daily average temperature with 3-day rolling range

Let's visualize it so that it's easy to highlight and identify trends in the data:

```
(
    daily_avg_temperature_lf
    .with_columns(
        pl.col('daily_avg_temp').rolling_map(get_range, window_
size=3).alias('3day_rolling_range')
    )
    .collect()
    .plot.line(
        x='date',
        y=['daily_avg_temp', '3day_rolling_range'],
        color=['skyblue', 'gray'],
        width=800,
        height=400
    )
    .opts(legend_position='bottom_right')
)
```

The preceding code will return the following output:

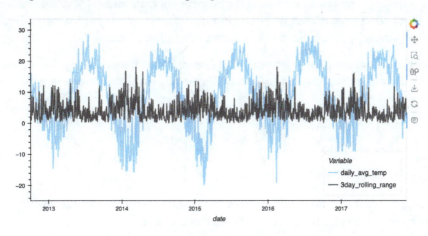

Figure 9.16 – A line plot showing the daily average temperature and 3-day rolling range

Note

Just like the `.map_element()` method, the `.rolling_map()` method is slow. Only use it when you cannot implement your logic with built-in methods/expressions.

See also

Here are additional resources you can utilize to learn more about rolling calculations in Polars:

- `https://docs.pola.rs/py-polars/html/reference/expressions/computation.html`
- `https://docs.pola.rs/py-polars/html/reference/expressions/api/polars.Expr.rolling_mean.html`
- `https://docs.pola.rs/py-polars/html/reference/expressions/api/polars.Expr.rolling.html`
- `https://docs.pola.rs/py-polars/html/reference/lazyframe/api/polars.LazyFrame.rolling.html`
- `https://docs.pola.rs/py-polars/html/reference/expressions/api/polars.Expr.rolling_map.html`

Resampling techniques

In time series analysis, there are two types of resampling techniques, upsampling and downsampling. Resampling in the context of time series analysis means changing the frequency of the data points.

Upsampling refers to the process of increasing the frequency or granularity of your data. This means converting a time series from a lower frequency to a higher frequency. If you have week-level data, then upsampling could be used to fill in values to create daily data points.

On the other hand, downsampling is the process of decreasing the frequency or granularity of your data. If you have day-level data, then this may involve aggregating your data to create monthly intervals.

In this recipe, we'll look at how to apply both upsampling and downsampling in Python Polars.

How to do it...

Here's how to apply resampling techniques.

1. The Toronto weather dataset contains hour-level data. Let's see how we can downsample the data from hourly to weekly. We'll look at humidity this time:

```
(
    lf
    .set_sorted('datetime')
    .group_by_dynamic(
        'datetime', every='1w'
    )
```

```
    .agg(pl.col('humidity').mean().round(1))
    .head(10)
    .collect()
)
```

The preceding code will return the following output:

shape: (10, 2)

datetime	humidity
datetime[µs]	f64
2012-10-01 00:00:00	63.1
2012-10-08 00:00:00	62.1
2012-10-15 00:00:00	76.1
2012-10-22 00:00:00	70.0
2012-10-29 00:00:00	80.0
2012-11-05 00:00:00	68.7
2012-11-12 00:00:00	68.5
2012-11-19 00:00:00	81.0
2012-11-26 00:00:00	69.0
2012-12-03 00:00:00	82.4

Figure 9.17 – Avg humidity at hourly frequency

Visualizing the result in a line plot is done as follows:

```
from datetime import datetime

(
    lf
    .set_sorted('datetime')
    .group_by_dynamic(
        'datetime', every='1w'
    )
    .agg(pl.col('humidity').mean().round(1))
    .filter(
        pl.col('datetime').dt.date().is_between(
            datetime(2012,10,1),
            datetime(2012,10,31)
        )
    )
)
```

```
        .collect()
        .plot.line(
            x='datetime',
            y='humidity',
            color=['skyblue'],
            width=1000,
            height=400
        )
    )
```

The preceding code will return the following output:

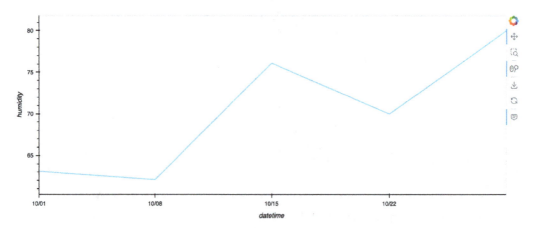

Figure 9.18 - A line plot showing the avg humidity at hourly frequency

We can see an upward trend in humidity towards the end of the month.

2.   Now we'll upsample the data from hourly to every 30 minutes:

```
upsampled_df = (
    lf
    .set_sorted('datetime')
    .collect()
    .upsample(
        time_column='datetime',
        every='30m',
        maintain_order=True
    )
    .select(
        'datetime',
```

```
        pl.col('humidity')
    )
)
upsampled_df.head(10)
```

The preceding code will return the following output:

shape: (10, 2)

datetime	humidity
datetime[μs]	f64
2012-10-01 12:00:00	null
2012-10-01 12:30:00	null
2012-10-01 13:00:00	82.0
2012-10-01 13:30:00	null
2012-10-01 14:00:00	81.0
2012-10-01 14:30:00	null
2012-10-01 15:00:00	79.0
2012-10-01 15:30:00	null
2012-10-01 16:00:00	77.0
2012-10-01 16:30:00	null

Figure 9.19 – The upsampled data with 30-min frequency

Notice that the `.upsample()` method created the new rows with null values for the humidity column.

From here, you can supplement those null values with whatever fits your needs. In this example, we'll use linear interpolation to fill null values:

```
(
    upsampled_df
    .with_columns(
        pl.col('humidity').interpolate()
    )
    .head(10)
)
```

The preceding code will return the following output:

shape: (10, 2)

datetime	humidity
datetime[µs]	f64
2012-10-01 12:00:00	null
2012-10-01 12:30:00	null
2012-10-01 13:00:00	82.0
2012-10-01 13:30:00	81.5
2012-10-01 14:00:00	81.0
2012-10-01 14:30:00	80.0
2012-10-01 15:00:00	79.0
2012-10-01 15:30:00	78.0
2012-10-01 16:00:00	77.0
2012-10-01 16:30:00	76.5

Figure 9.20 – Null values filled with interpolation

Visualizing the output in an area plot, limiting the date range to only a month, is done as follows:

```
(
    upsampled_df
    .with_columns(
        pl.col('humidity').interpolate()
    )
    .filter(
        pl.col('datetime').dt.date().is_between(
            datetime(2012,10,1),
            datetime(2012,10,31)
        )
    )
    .plot.area(
        x='datetime',
        y='humidity',
        color=['skyblue'],
        width=1000,
        height=400,
        alpha=0.5
    )
)
```

The preceding code will return the following output:

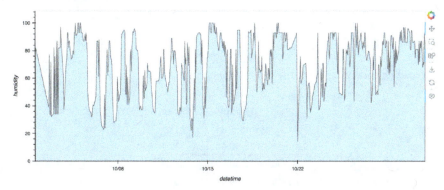

Figure 9.21 – An area plot showing the avg humidity with a 30-minute frequency

The same pattern in *Figure 9.18* is found in this visualization; namely that closer to the end of October 2012, less variance is seen. The average is also higher.

3. Given that the `.upsample()` method fills gaps or holes in your time values, you can use it for that purpose only without increasing the frequency of the data.

    I.   First, create a LazyFrame with an inconsistent intervals:

    ```
    datetime_with_gaps_lf = (
        lf.filter(
            ~pl.col('datetime').dt.hour().is_in([13,15,16,19])
        )
    )
    ```

    II. Second, apply `.upsample()` with the hourly interval, which is the original frequency in the dataset. This is only to fill gaps in the data:

    ```
    (
        datetime_with_gaps_lf
        .set_sorted('datetime')
        .collect()
        .upsample(
            time_column='datetime',
            every='1h',
            maintain_order=True
        )
        .select(
            'datetime',
    ```

```
                pl.col('humidity')
        )
    .head(10)
)
```

The preceding code will return the following output:

shape: (10, 2)

datetime	humidity
datetime[µs]	f64
2012-10-01 12:00:00	null
2012-10-01 13:00:00	null
2012-10-01 14:00:00	81.0
2012-10-01 15:00:00	null
2012-10-01 16:00:00	null
2012-10-01 17:00:00	76.0
2012-10-01 18:00:00	74.0
2012-10-01 19:00:00	null
2012-10-01 20:00:00	70.0
2012-10-01 21:00:00	69.0

Figure 9.22 – The data with gaps filled

4. Since the `.upsample()` method is only available in DataFrames, here's a workaround to accomplish the same thing with your data still being in a LazyFrame:

```
datetime_range_lf = pl.LazyFrame({
    'datetime': pl.datetime_range(
        start=lf.select(pl.col('datetime').min()).collect()
[0,0],
        end=lf.select(pl.col('datetime').max()).collect()[0,0],
        interval='30m',
        eager=True
    )
})

(
    datetime_range_lf
    .join(lf, on='datetime', how='left', coalesce=True)
    .select(
        'datetime',
```

```
            pl.col('humidity')
    )
    .collect()
    .head(10)
)
```

The preceding code will return the same output as in *Figure 9.19*.

How it works...

The `.group_by_dynamic()` method is really powerful for downsampling. It lets you adjust several useful parameters, including but not limited to `every`, `period`, `offset`, `closed`, and `start_by`. The `every` parameter helps specify the interval of the window. The `period` parameter lets you define the length of the window. The `offset` and `start_by` parameters change the start position of the window. The `closed` parameter helps you determine which sides of the interval should be inclusive.

The difference between the `.group_by()` method and the `.group_by_dynamic()` method is that the latter allows a row to be a member of multiple groups. As mentioned previously, those parameters help adjust how the temporal window should behave with the `.group_by_dynamic()`. On the other hand, the `.group_by()` method simply summarizes or aggregates data.

The `.upsample()` method is for upsampling because it can increase the frequency of your data. Also, it creates a new row per interval if it doesn't exist. In our example, we increased the frequency from hourly to every 30 minutes. Each new row was inserted with a null value. You can also offset the start of the range with the `offset` parameter. The `by` parameter helps group the data based on the specified columns initially, and then upsample for each individual group.

One downside for `.upsample()` is that it's only available for DataFrames. If you want to keep your LazyFrame, the approach with the `pl.datetime_range()` function is the way to do it.

An integer index column can also be used as its time column with these resampling methods in Polars. Parameters such as `every`, `period`, and `offset` allow you to specify the time with a string language such as "*1h*" and "*30m*", but you can also use something like "*1i*", which is 1 index count or length of 1.

> **Important**
>
> Both of these resampling methods, `.group_by_dynamic()` and `.upsample()`, require the time column to be sorted. That can be done either with the `.set_sorted()` method if the column was already sorted, or the `.sort()` method.

> **Note**
>
> You might notice that the `.rolling()` DataFrame/LazyFrame method introduced in the previous recipe is similar to the `.group_by_dynamic()` method. Here's what the Polars documentation says about the former: "The windows are now determined by the individual values and are not of constant intervals. For constant intervals use `.group_by_dynamic()`."
>
> Although these may appear related or similar in function because they both group data by the time column, I try not to think of these two methods in the same bucket. Just think of the `.rolling()` method for rolling calculations and the `.group_by_dynamic()` method for downsampling.

See also

Here are additional resources that may be helpful for learning more about resampling techniques in Polars:

- `https://docs.pola.rs/py-polars/html/reference/lazyframe/api/polars.LazyFrame.group_by_dynamic.html`

- `https://docs.pola.rs/py-polars/html/reference/dataframe/api/polars.DataFrame.upsample.html`

- `https://docs.pola.rs/py-polars/html/reference/expressions/api/polars.datetime_range.html`

- `https://www.rhosignal.com/posts/filling-gaps-lazy-mode/`

Time series forecasting with the functime library

Time series forecasting is a form of predictive analytics for time series data. It is to predict the future values based on historical data using statistical models. `functime` is a machine learning library for time series forecasting and feature extraction in Polars. It enables you to build time series forecasting models utilizing the Polars speed. `functime` is the Polars' version of `tsfresh`, the popular time series feature extraction library.

In this recipe, we'll cover how to build a simple time series forecasting model with the `functime` library, including feature extraction and plotting.

> **Important**
>
> As of the time of writing, Polars just upgraded to version 1.0.0 and `functime` has some compatibility issues with it. You may encounter an error after *step 2*; however, you can run those later steps with Polars version 0.20.31 (you'll also need to change `lf.collect_schema().names()` to `lf.columns` for the code in *step 1*). If `functime` resolved the compatibility issues by the time you're reading this, you can run all the steps in this recipe without issues.

Getting ready

Install the `functime` library with pip:

```
pip install functime
```

Note that `functime` has several dependencies including `FLAML`, `scikit-learn`, `scipy`, `cloudpickle`, `holidays`, and `joblib`.

In this recipe, we'll use a different dataset based on the same data source, containing temperature data from multiple cities such as Toronto, New York, and Seattle. The structure of the dataset stays the same and the only difference is that this dataset will have temperatures per city and datetime columns. This kind of structure is called *panel* data, involving multiple time series data stacked on top of each other. It consists of multiple entity columns (`city` and `datetime`) and observed values (temperature). `functime` is designed to build forecasting models on panel data.

Reading the data in a LazyFrame is done as follows:

```
lf = pl.scan_csv('../data/historical_temperatures.csv', try_parse_
dates=True)

lf.head().collect()
```

The preceding code will return the following output:

shape: (5, 3)

datetime	city	temperature
datetime[μs]	str	f64
2012-10-01 12:00:00	"Toronto"	null
2012-10-01 13:00:00	"Toronto"	286.26
2012-10-01 14:00:00	"Toronto"	286.262541
2012-10-01 15:00:00	"Toronto"	286.269518
2012-10-01 16:00:00	"Toronto"	286.276496

Figure 9.23 – The first five rows in the panel data

Let's check the values in the `city` column:

```
lf.select('city').unique().sort('city').collect()
```

The preceding code will return the following output:

shape: (6, 1)

city
str
"San Francisco"
"Vancouver"
"New York"
"Seattle"
"Las Vegas"
"Toronto"

Figure 9.24 – Unique cities in the city column

You can also use `.group_by()` in combination with `.head()` or `.tail()` to check the first *n* rows for every value in the group-by column:

```
lf.group_by('city').head(3).collect()
```

The preceding code will return the following output:

shape: (18, 3)

city	datetime	temperature
str	datetime[µs]	f64
"Vancouver"	2012-10-01 12:00:00	null
"Vancouver"	2012-10-01 13:00:00	284.63
"Vancouver"	2012-10-01 14:00:00	284.629041
"New York"	2012-10-01 12:00:00	null
"New York"	2012-10-01 13:00:00	288.22
"New York"	2012-10-01 14:00:00	288.247676
"San Francisco"	2012-10-01 12:00:00	null
"San Francisco"	2012-10-01 13:00:00	289.48
"San Francisco"	2012-10-01 14:00:00	289.474993
"Las Vegas"	2012-10-01 12:00:00	null
"Las Vegas"	2012-10-01 13:00:00	293.41
"Las Vegas"	2012-10-01 14:00:00	293.403141
"Seattle"	2012-10-01 12:00:00	null
"Seattle"	2012-10-01 13:00:00	281.8
"Seattle"	2012-10-01 14:00:00	281.797217
"Toronto"	2012-10-01 12:00:00	null
"Toronto"	2012-10-01 13:00:00	286.26
"Toronto"	2012-10-01 14:00:00	286.262541

Figure 9.25 – The first n rows for every value in the group-by column (city)

How to do it...

Here's how to build a forecasting model with `functime`.

1. Prepare the data at the month level as well as converting the temperature from Kelvin to Celsius:

```python
time_col, entity_col, value_col = lf.collect_schema().names()


y = (
    lf
    .group_by_dynamic(
        time_col,
        every='1mo',
        group_by=entity_col,
    )
    .agg(
        (pl.col('temperature').mean()-273.15).round(1),
    )
)


y.group_by('city').head(3).collect()
```

The preceding code will return the following output:

shape: (18, 3)

city	datetime	temperature
str	datetime[µs]	f64
"Toronto"	2012-10-01 00:00:00	10.3
"Toronto"	2012-11-01 00:00:00	4.3
"Toronto"	2012-12-01 00:00:00	1.1
"New York"	2012-10-01 00:00:00	14.4
"New York"	2012-11-01 00:00:00	5.9
"New York"	2012-12-01 00:00:00	4.7
"Seattle"	2012-10-01 00:00:00	11.2
"Seattle"	2012-11-01 00:00:00	7.9
"Seattle"	2012-12-01 00:00:00	5.2
"San Francisco"	2012-10-01 00:00:00	16.5
"San Francisco"	2012-11-01 00:00:00	13.4
"San Francisco"	2012-12-01 00:00:00	10.3
"Las Vegas"	2012-10-01 00:00:00	20.9
"Las Vegas"	2012-11-01 00:00:00	14.4
"Las Vegas"	2012-12-01 00:00:00	9.0
"Vancouver"	2012-10-01 00:00:00	10.1
"Vancouver"	2012-11-01 00:00:00	7.2
"Vancouver"	2012-12-01 00:00:00	4.4

Figure 9.26 – The first n rows for every city with datetime and temperature converted

1. Split the dataset into training and testing sets. functime has similar syntax to scikit-learn for the train-test split:

```
def create_train_test_sets(
    y,
    entity_col,
    time_col,
    test_size):
    from functime.cross_validation import train_test_split

    X = y.select(entity_col, time_col)
    y_train, y_test = (
        y
        .select(entity_col, time_col, value_col)
        .pipe(train_test_split(test_size))
    )
    X_train, X_test = X.pipe(train_test_split(test_size))

    return X_train, X_test, y_train, y_test

test_size = 3
X_train, X_test, y_train, y_test = create_train_test_sets(y,
entity_col, time_col, test_size)
```

2. Predict the values and calculate the **Mean Absolute Scaled Error (MASE)** metric for the model:

```
def predict_with_linear_model(
    lags,
    freq,
    y_train,
    fh
):
    from functime.forecasting import linear_model

    forecaster = linear_model(lags=lags, freq=freq)
    forecaster.fit(y=y_train)
    y_pred = forecaster.predict(fh=fh)
    return y_pred

y_pred = predict_with_linear_model(24, '1mo', y_train, test_
size)
```

```
from functime.metrics import mase
scores = mase(y_true=y_test, y_pred=y_pred, y_train=y_train)
display(y_pred, scores)
```

The preceding code will return the following output:

shape: (18, 3)

city	datetime	temperature
str	datetime[μs]	f64
"Toronto"	2017-09-01 00:00:00	21.900375
"Toronto"	2017-10-01 00:00:00	19.132336
"Toronto"	2017-11-01 00:00:00	12.42299
"Las Vegas"	2017-09-01 00:00:00	33.356968
"Las Vegas"	2017-10-01 00:00:00	26.222855
"Las Vegas"	2017-11-01 00:00:00	19.330391
"New York"	2017-09-01 00:00:00	24.757912
"New York"	2017-10-01 00:00:00	22.249531
"New York"	2017-11-01 00:00:00	14.419162
"Seattle"	2017-09-01 00:00:00	16.692822
"Seattle"	2017-10-01 00:00:00	14.990096
"Seattle"	2017-11-01 00:00:00	12.816636
"San Francisco"	2017-09-01 00:00:00	19.970325
"San Francisco"	2017-10-01 00:00:00	18.315979
"San Francisco"	2017-11-01 00:00:00	16.368904
"Vancouver"	2017-09-01 00:00:00	16.471996
"Vancouver"	2017-10-01 00:00:00	14.267774
"Vancouver"	2017-11-01 00:00:00	11.530054

Figure 9.27 – Forecasted/predicated values

And here's the MASE values:

shape: (6, 2)

city	mase
str	f64
"Seattle"	1.286418
"New York"	1.082173
"Toronto"	1.341056
"Vancouver"	0.737572
"San Francisco"	0.542017
"Las Vegas"	1.151459

Figure 9.28 – MASE values for each city

3. Visualizing the predicted values and the original data using `hvplot` is done as follows:

```
actual_viz = (
    y
```

```
        .collect()
        .plot.line(
            x='datetime', y='temperature', by='city', subplots=True
        )
        .cols(2)
)

pred_viz = (
    y_pred
    .plot.line(
        x='datetime', y='temperature', by='city', subplots=True
    )
    .cols(2)
)

actual_viz * pred_viz
```

The preceding code will return the following output:

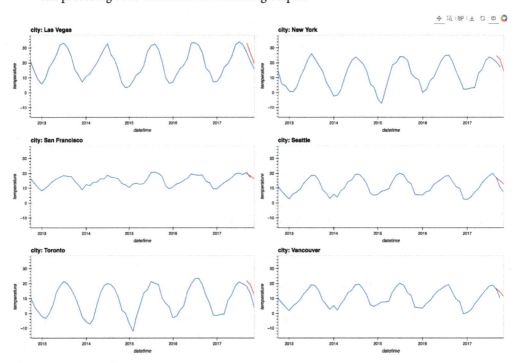

Figure 9.29 - Original values and predicated values visualized in line charts

How it works...

As you saw, building a forecasting model with `functime` is not complicated at all, especially if you have some experience with other forecasting libraries. The process of creating training and testing sets looks very much like how you do it in scikit-learn.

You might've noticed the use of the `.pipe()` method in *step 2*. It can take **user-defined functions (UDFs)** in sequence so that your code is more structured.

For feature extraction, `functime` provides the custom `ts` namespace that contains every feature. The `ts` namespace can be used just like other namespaces such as `str` and `list`. They allow you to access useful methods in any Polars expression. We'll see an example of extracting features using the `ts` namespace shortly.

There is more...

Here's an example of how to extract features in `functime`:

```python
from functime.seasonality import add_calendar_effects

y_features = (
    lf
    .group_by_dynamic(
        time_col,
        every='1mo',
        by=entity_col,
    )
    .agg(
        (pl.col('temperature').mean()-273.15).round(1),
        pl.col(value_col).ts.binned_entropy(bin_count=10)
        .alias('binned_entropy'),
        pl.col(value_col).ts.lempel_ziv_complexity(threshold=3)
        .alias('lempel_ziv_complexity'),
        pl.col(value_col).ts.longest_streak_above_mean()
        .alias('longest_streak_above_mean')
    )
    .pipe(add_calendar_effects(['month']))
)

y_features.head().collect()
```

The preceding code will return the following output:

shape: (5, 7)

city	datetime	temperature	binned_entropy	lempel_ziv_complexity	longest_streak_above_mean	month
str	datetime[µs]	f64	f64	f64	u64	cat
"Toronto"	2012-10-01 00:00:00	10.3	2.02934	0.051913	94	"10"
"Toronto"	2012-11-01 00:00:00	4.3	1.969688	0.051389	87	"11"
"Toronto"	2012-12-01 00:00:00	1.1	2.10535	0.051075	87	"12"
"Toronto"	2013-01-01 00:00:00	-2.1	2.082969	0.051075	171	"1"
"Toronto"	2013-02-01 00:00:00	-3.4	1.946257	0.053571	130	"2"

Figure 9.30 – Data with extracted features

In the preceding code, note that the `add_calendar_effects()` function adds a month column. The seasonality API provides functions to extract calendar/holiday effects and Fourier features for time series seasonality.

See also

Please refer to the `functime` documentation to learn more:

- https://docs.functime.ai/user-guide/forecasting/

- https://docs.functime.ai/user-guide/feature-extraction/

- https://docs.functime.ai/user-guide/seasonality/

10

Interoperability with Other Python Libraries

Although Polars is an awesome tool that's fast and efficient, there are times when interoperating with other tools or libraries is crucial in your data projects. The good news is that there are libraries out there already that can work with Polars. In *Chapter 9, Time Series Analysis*, you've already seen that it works well with the `functime` and `plotly` libraries. Polars can also work with other Python libraries such as pandas, NumPy, PyArrow, and DuckDB to name a few. As Polars matures more as a tool, there will be more libraries and tools, making the integration and interoperability between Polars and the whole Python data ecosystem more seamless. For instance, having a seamless integration with other Python libraries benefits Polars by providing functionalities it doesn't yet have. It'll give you more options for how you implement your solution.

By the end of this chapter, you'll gain an understanding of how Polars interoperates with pandas, NumPy, PyArrow, and DuckDB.

In this chapter we're going to cover the following main topics:

- Converting to and from a pandas DataFrame
- Converting to and from NumPy arrays
- Interoperating with PyArrow
- Integrating with DuckDB

Technical requirements

You can download the datasets and code in the GitHub repository:

- Datasets: `https://github.com/PacktPublishing/Polars-Cookbook/tree/main/data`
- Code: `https://github.com/PacktPublishing/Polars-Cookbook/tree/main/Chapter10`

It is assumed that you have installed the Polars library in your Python environment:

```
>>> pip install polars
```

It is also assumed that you imported it in your code:

```
import polars as pl
```

Execute the following code to install the other libraries:

```
pip install pandas numpy pyarrow duckdb
```

Converting to and from a pandas DataFrame

Many of you have used pandas before, especially in your day-to-day work. Although pandas and Polars are often compared as one-or-the-other tools, you can use these tools to supplement each other. Polars allows you to convert between pandas and Polars DataFrames, which is exactly what we'll cover in this recipe.

Getting ready

You need `pandas` and `pyarrow` installed for this recipe to work. Execute the following code to make sure that you have them installed:

```
pip install pandas pyarrow
```

How to do it...

Here's how to convert to and from pandas DataFrames. We'll first create a Polars DataFrame and then go through ways to convert back and forth between Polars and pandas:

1. Create a Polars DataFrame from a Python dictionary:

```
df = pl.DataFrame({
    'a': [1,2,3],
    'b': [4,5,6]
```

```
})
type(df)
```

The preceding code will return the following output:

```
>> polars.dataframe.frame.DataFrame
```

2. Convert a Polars DataFrame to a pandas DataFrame using the .to_pandas() method:

```
pandas_df = df.to_pandas()
type(pandas_df)
```

The preceding code will return the following output:

```
>> pandas.core.frame.DataFrame
```

3. Convert a pandas DataFrame to a Polars DataFrame using the .from_pandas() method:

```
df = pl.from_pandas(pandas_df)
type(df)
```

The preceding code will return the following output:

```
>> polars.dataframe.frame.DataFrame
```

4. If you want to allow zero copy operations, then you need to enable the use_pyarrow_extension_array parameter:

```
df.to_pandas(use_pyarrow_extension_array=True).dtypes
```

The preceding code will return the following output:

```
>>
a int64[pyarrow]
b int64[pyarrow]
dtype: object
```

You can also create a Polars DataFrame by wrapping a pandas DataFrame using pl.DataFrame():

```
type(pl.DataFrame(pandas_df))
```

The preceding code will return the following output:

```
>> polars.dataframe.frame.DataFrame
```

How it works...

Polars has built-in methods to interoperate with pandas such as .from_pandas() and .to_pandas(). Each method is descriptive enough that you can see that .from_pandas() is used for reading data into Polars from pandas, whereas .to_pandas() is used to convert Polars objects into pandas.

The `use_pyarrow_extension_array` parameter of the `.to_pandas()` method uses PyArrow-supported arrays instead of NumPy arrays for the columns within the pandas DataFrame. This enables zero-copy operations and maintains the integrity of null values.

There's more...

You can convert to and from a pandas Series to a Polars Series:

```
s = pl.Series([1,2,3])
type(s.to_pandas())
```

The preceding code produces the following:

```
>> pandas.core.series.Series
```

The `.from_pandas()` method returns a Series object when a pandas Series was passed in:

```
type(pl.from_pandas(s.to_pandas()))
```

The preceding code produces the following:

```
>> polars.series.series.Series
```

See also

Please refer to these Polars documentation pages to learn more about reading from and converting to a pandas DataFrame:

- `https://docs.pola.rs/py-polars/html/reference/dataframe/api/polars.DataFrame.to_pandas.html`
- `https://docs.pola.rs/py-polars/html/reference/api/polars.from_pandas.html`
- `https://docs.pola.rs/py-polars/html/reference/dataframe/index.html`

Converting to and from NumPy arrays

NumPy can also work with Polars. You can convert back and forth between Polars and NumPy. Also, in cases where Polars lacks a particular function, we have the option to utilize NumPy instead while benefiting from efficient columnar operations via the NumPy API.

In this recipe, we'll cover how to convert from and to NumPy arrays, as well as how to use NumPy functions directly in Polars.

Getting ready

You need NumPy installed for this recipe:

```
pip install numpy
```

How to do it...

Here's how you can utilize NumPy in Polars. We'll first create a NumPy array and then convert back and forth between Polars and NumPy.

1. Create a NumPy array:

    ```
    arr = np.array([[1,2,3], [4,5,6]])
    ```

2. Create a Polars DataFrame from the NumPy array:

    ```
    df = pl.from_numpy(arr, schema=['a', 'b'], orient='col')
    df
    ```

    The preceding code will return the following output:

    shape: (3, 2)

a	b
i64	i64
1	4
2	5
3	6

    Figure 10.1 – A Polars DataFrame created from a NumPy array

3. Convert a Polars DataFrame to a NumPy array:

    ```
    df.to_numpy()
    ```

    The preceding code will return the following output:

    ```
    >> array([[1, 4], [2, 5], [3, 6]])
    ```

4. Setting the `structured` parameter to `True` preserves the column data, such as the `data` type:

    ```
    df.to_numpy(structured=True)
    ```

    The preceding code will return the following output:

    ```
    >> array([(1, 4), (2, 5), (3, 6)], dtype=[('a', '<i8'), ('b',
    '<i8')])
    ```

5.   Use a NumPy function on a Polars column expression:

```
(
    df
    .with_columns(
        np.gcd(pl.col('a'), pl.col('b')).alias('gcd')
    )
)
```

The preceding code will return the following output:

shape: (3, 3)

a	b	gcd
i64	i64	i64
1	4	1
2	5	1
3	6	3

Figure 10.2 – A Polars DataFrame with a new column created from a NumPy function

How it works...

You can not only convert back and forth between Polars DataFrames and NumPy arrays but also use NumPy functions directly on Polars column expressions. This capability can significantly enhance the range of tasks achievable with Polars.

In the preceding steps, we created a NumPy array and converted back and forth between Polars and NumPy. In *step 5*, we used np.gcd() to calculate the greatest common divisor of the a and b columns. That's just one example of how you can utilize NumPy to provide more functionality in Polars.

You can refer to this NumPy documentation page to see which NumPy functions are supported in Polars:

`https://numpy.org/doc/stable/reference/ufuncs.html#available-ufuncs`

There's more...

You can also convert a Polars Series to a NumPy array:

```
s = pl.Series([1,2,3])
s.to_numpy()
```

The preceding code will produce the following output:

```
>> array([1, 2, 3])
```

> **Note**
>
> You can convert a 1D NumPy array to Polars as well, but the output will be a DataFrame with a column.

See also

Refer to the Polars documentation that follows to learn more about utilizing NumPy in Polars:

- `https://docs.pola.rs/py-polars/html/reference/api/polars.from_numpy.html`

- `https://docs.pola.rs/py-polars/html/reference/dataframe/api/polars.DataFrame.to_numpy.html`

- `https://docs.pola.rs/user-guide/expressions/numpy/`

Interoperating with PyArrow

Apache Arrow serves as a language-independent columnar memory format. It encompasses a range of technologies that empower big data systems to efficiently store, process, and transfer data. The PyArrow library is the Python API of Apache Arrow. Polars uses Apache Arrow's columnar format as its memory model. Just like pandas uses NumPy for its in-memory representation of data, Polars uses PyArrow (since pandas version 2.0, it has had an added functionality to use PyArrow as its in-memory format).

The interoperability between PyArrow and Polars is great because you can not only convert back and forth between Polars DataFrames and PyArrow datasets but also use PyArrow with other aspects of things such as reading and writing files, as you saw in *Chapter 2, Reading and Writing Files*.

In this recipe, we'll look at how to convert back and forth between Polars DataFrames and PyArrow datasets, as well as directly reading from and writing to PyArrow datasets and tables.

Getting ready

We'll be using the Parquet file that we used in *Chapter 2*: `https://github.com/PacktPublishing/Polars-Cookbook/tree/main/data/venture_funding_deals_partitioned`

How to do it...

Here's how to utilize PyArrow in Polars. We will start by reading the Parquet file into a PyArrow dataset. Then we'll demonstrate how we can work with PyArrow in Polars:

1. Read the partitioned Parquet file into a PyArrow dataset:

    ```
    import pyarrow.dataset as ds

    file_path = '../data/venture_funding_deals_partitioned'
    part = ds.partitioning(flavor='hive')
    dataset = ds.dataset(file_path, partitioning=part)

    dataset.head(5)
    ```

    The preceding code will return the following output:

    ```
    >>
    pyarrow.Table
    Company: large_string
    Amount: large_string
    Lead investors: large_string
    Valuation: large_string
    Date reported: large_string
    Industry: string
    ----
    Company: [["Restaurant365"],["Madhive"],...,
    ["Indigo"],["Chronosphere"]]
    Amount: [["$135,000,000"],["$300,000,000"],...,
    ["$250,000,000"],["$115,000,000"]]
    Lead investors: [["KKR, L Catterton"],
    ["Goldman Sachs Asset Management"],...,
    ["Flagship Pioneering, State of Michigan Retirement System,
    Lingotto"],["GV"]]
    Valuation: [["$1,000,000,000"],["$1,000,000,000"]
    ,...,["na"],["n/a"]]
    Date reported: [["5/19/23"],["6/13/23"]
    ,...,["9/15/23"],["1/9/23"]]
    Industry: [["Accounting"],["Advertising"]
    ,...,["Agriculture"],["Analytics"]]
    ```

2. Convert the PyArrow dataset to a PyArrow table. Then convert it to a Polars DataFrame:

    ```
    df = pl.from_arrow(dataset.to_table())
    ```

```
df.head()
```

The preceding code will return the following output:

shape: (5, 6)

Company	Amount	Lead investors	Valuation	Date reported	Industry
str	str	str	str	str	str
"Restaurant365"	"$135,000,000"	"KKR, L Cattert...	"$1,000,000,000...	"5/19/23"	"Accounting"
"Madhive"	"$300,000,000"	"Goldman Sachs ...	"$1,000,000,000...	"6/13/23"	"Advertising"
"Ursa Major,"	"$100,000,000"	"BlackRock, Spa...	"n/a"	"4/26/23"	"Aerospace"
"Indigo"	"$250,000,000"	"Flagship Pione...	"na"	"9/15/23"	"Agriculture"
"Chronosphere"	"$115,000,000"	"GV"	"n/a"	"1/9/23"	"Analytics"

Figure 10.3 – The DataFrame read from a PyArrow table

You can always convert a Polars DataFrame to a PyArrow table:

```
df.to_arrow()
```

The preceding code will return the following output:

```
>>
pyarrow.Table
Company: large_string
Amount: large_string
Lead investors: large_string
Valuation: large_string
Date reported: large_string
Industry: large_string
----
Company: [["Restaurant365","Madhive","Ursa
Major,","Indigo","Chronosphere",...,"Professional
Fighters League","Newlight Technologies","Pivotal
Commware","Via","Aquaback Technologies"]]
Amount: [["$135,000,000","$300,000,000","$100,000,000","$250,000
,000","$115,000,000",...,"$100,000,000","$125,000,000","$102,000
,000","$110,000,000","$110,000,000"]]
Lead investors: [["KKR, L Catterton","Goldman Sachs Asset
Management","BlackRock, Space Capital","Flagship Pioneering,
State of Michigan Retirement System, Lingotto","GV",...,"SRJ
Sports Investments","GenZero","Gates Frontier, Tracker
Capital","83North","Global Emerging Markets Group"]]
Valuation: [["$1,000,000,000","$1,000,000,000","n/a","na","n/
a",...,"n/a","n/a","n/a","$3,500,000,000","n/a"]]
Date reported: [["5/19/23","6/13/23","4/26/23","9/15/23","1/9/23
",...,"8/30/23","8/3/23","8/17/23","2/13/23","6/28/23"]]
Industry:
[["Accounting","Advertising","Aerospace",
```

```
"Agriculture","Analytics",...,
"Sports","Sustainability","Telecommunications"
,"Transportation","Water"]]
```

We can also scan a PyArrow dataset into a Polars LazyFrame:

```
lf = pl.scan_pyarrow_dataset(dataset)
lf.head().collect()
```

The preceding code will return the same output as in *Figure 10.3*.

As you may have figured, you can convert a Polars Series to an Arrow array:

```
(
    lf
    .select('Company')
    .collect()
    .to_series()
    .to_arrow()
)
```

The preceding code will return the following output:

```
>>
<pyarrow.lib.LargeStringArray object at 0x12fd3d9c0>
[
  "Restaurant365",
  "Madhive",
  "Ursa Major,",
  "Indigo",
  "Chronosphere",
  "AlphaSense",
  "Skims",
  "SandboxAQ",
  "Humane",
  "OpenAI",
  ...
  "Eagle Eye Networks",
  "Enfabrica",
  "Axiom Space",
  "Sierra Space",
  "Astranis",
  "Professional Fighters League",
  "Newlight Technologies",
  "Pivotal Commware",
  "Via",
  "Aquaback Technologies"
]
```

How it works...

Both the `pl.from_arrow()` function and the `.to_arrow()` method help convert from and to an Arrow table or array. The `pl.scan_pyarrow_dataset()` is useful when you utilize lazy mode by scanning the data instead of eagerly reading it. Note that it may not utilize all the optimizations that are available in the full Polars API. For example, it can only push down predicates that are allowed by PyArrow.

> **Note**
>
> If you're interested in learning about reading IPC/Feather file format, please refer to *Chapter 2, Reading and Writing Files.*

See also

Please refer to these additional resources to learn more about utilizing PyArrow in Polars:

- `https://docs.pola.rs/py-polars/html/reference/api/polars.from_arrow.html`

- `https://docs.pola.rs/py-polars/html/reference/dataframe/api/polars.DataFrame.to_arrow.html`

- `https://docs.pola.rs/py-polars/html/reference/api/polars.scan_pyarrow_dataset.html`

- `https://docs.pola.rs/api/python/stable/reference/series/api/polars.Series.to_arrow.html#polars.Series.to_arrow`

- `https://arrow.apache.org/docs/python/api.html`

Integrating with DuckDB

DuckDB is an in-process analytical database. The speed is like that of Polars, which is really fast. Like Polars, DuckDB has built-in integrations with other tools. One example is its integration with Arrow. Just like Polars, DuckDB allows for zero-copy to and from the Arrow format. DuckDB has APIs in many languages such as Python, C, Go, R, and Swift. You can check out DuckDB's documentation to learn more about it at `https://duckdb.org/`.

DuckDB has a solid SQL API with a rich set of features. It's superior to the SQL API in Polars as of the time of writing this. You can also run a SQL query on a Polars DataFrame directly without copying data. Even if you convert a DuckDB relation to a Polars DataFrame or vice versa, it allows you to enable zero-copy operations, thanks to Apache Arrow columnar memory format used both in Polars and DuckDB.

Getting ready

You need to have PyArrow installed as Arrow is the fundamental technology through which DuckDB and Polars can interoperate. If you haven't, execute the following code in the command line:

```
pip install duckdb
```

How to do it...

Here's how to use DuckDB with Polars. We'll first see how to run an SQL query on a Polars DataFrame with DuckDB. And then we'll see how to convert a DuckDB relation to a Polars DataFrame.

1. Run a SQL query with DuckDB directly on a Polars DataFrame:

    ```
    import duckdb

    df = pl.DataFrame({
        'a': [1,2,3]
    })
    rel = duckdb.sql('SELECT * FROM df')
    rel.show()
    ```

    The preceding code will return the following output:

    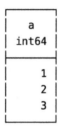

    Figure 10.4 – The result of a DuckDB SQL query on a Polars DataFrame

2. Convert a DuckDB relation to a Polars DataFrame:

    ```
    rel.pl()
    ```

The preceding code will return the following output:

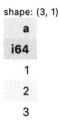

Figure 10.5 – A Polars DataFrame converted from a DuckDB relation

How it works...

In *step 1*, we ran a SQL query on a Polars DataFrame through DuckDB. In *step 2*, we converted a DuckDB relation to a Polars DataFrame. In both methods, we used DuckDB's functionalities. DuckDB is already compatible to work with Polars. This is all thanks to DuckDB having the zero-copy integration with Arrow, as well as Polars using Arrow for its memory model.

Although you can run DuckDB SQL on both Polars DataFrame and Polars LazyFrame, the resulting DuckDB relation doesn't retain the laziness from the input LazyFrame.

> **Tip**
> You might wonder which tool you should use (Polars or DuckDB) if you're solely using SQL. My answer is DuckDB, as the SQL API is more mature and includes the whole list of standard SQL functions, as well as more advanced ones such as those that work on nested data types. It's just a more full-featured SQL API than that of Polars as of the time of writing.

See also

To learn more about Polars integration with DuckDB or DuckDB in general, please refer to these additional resources:

- `https://duckdb.org/`
- `https://duckdb.org/docs/guides/python/polars.html`

11

Working with Common Cloud Data Sources

As more and more companies and organizations migrate their systems to the cloud, the ability to work with cloud data sources has become essential for any data processing tool. In *Chapter 2 Reading and Writing Files*, we've covered how we can read and write files in Polars. In this chapter, we are going to cover how we can work with common data sources hosted in the cloud.

The most challenging part is probably the configuration, not only for working with cloud sources but also anything in data projects. We'll be sure to cover how to do that as well. However, you get the most out of this chapter if you already have experience working with a cloud service, so that you may know how to troubleshoot access issues as they arise.

In this chapter, we're going to cover the following main topics:

- Working with Amazon S3
- Working with Azure Blob Storage
- Working with **Google Cloud Storage**
- Working with BigQuery
- Working with Snowflake

Note that it may incur costs if you create your account on any of the cloud providers, set up the services, and execute the code or steps specified in this chapter. Be mindful or thoughtful of what you're doing in your cloud account to avoid any surprise cloud bills.

> **Important**
>
> You'll see hard-coded credentials in some of the examples in this chapter, however, this is solely for demonstration purposes, and you should never hardcode your credentials directly in your code. You may store them in a separate file, for example, a `.env` file, so that it can be added to the `.gitignore` file for your credentials not to be shared around. Or you may choose to store them in environment variables or use a secret manager in the cloud platform of your choice.

> **Note**
>
> Some of the items we're going to create in each cloud service include naming object storage and tables. For those names, you may notice I use my initials, `yk`, as the suffix as they often need to be globally unique. I suggest that you either add your initials or name as the suffix.

Technical requirements

You can download the datasets and code in the GitHub repository:

- **Data**: https://github.com/PacktPublishing/Polars-Cookbook/tree/main/data
- **Code**: https://github.com/PacktPublishing/Polars-Cookbook/tree/main/Chapter11

It is assumed that you have installed the `polars` library in your Python environment:

```
>>> pip install polars
```

It is also assumed that you imported it into your code:

```
import polars as pl
```

You'd need to have **Amazon Web Services** (**AWS**), Azure, GCP, and Snowflake accounts to go through recipes in this chapter. The good news is that they offer free credits or a free trial, so you may not pay a dime to try out these platforms! They may even have always-free products. The following is the list of resources you can follow to create an account for each cloud vendor:

- **AWS**: https://aws.amazon.com/free
- **Azure**: https://azure.microsoft.com/en-us/free
- **GCP**: https://cloud.google.com/free
- **Snowflake**: https://signup.snowflake.com/

Working with Amazon S3

Amazon S3 is one of the most popular choices for an object storage service. Many organizations use it for a data lake, a data repository that stores both structured and unstructured data.

In this recipe, we'll look at how to read from and write to an S3 bucket in Polars.

Getting ready

You need your AWS account for this recipe. Follow the instructions in the link I mentioned under *Technical requirements* to create your AWS account.

Let's create an S3 bucket. I introduce how to do that in the following steps, but if you prefer reading the documentation on your own, here's a useful link: `https://docs.aws.amazon.com/AmazonS3/latest/userguide/creating-bucket.html`.

Here are the steps for creating an S3 bucket:

1. Navigate to `https://aws.amazon.com/` and sign in to your account.
2. You can either type `S3` in the search bar, click on **Services** on the top left, or just choose **S3** in **Recently visited** if you see one.

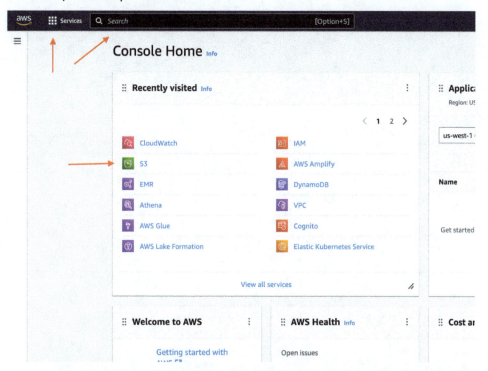

Figure 11.1 – AWS Console Home

3.   Once you navigated to your S3 buckets, click on **Create bucket**. You may see existing S3 buckets if you have any.

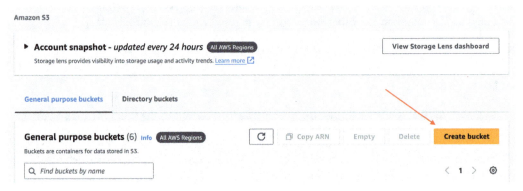

Figure 11.2 – S3 bucket list

4.   Choose your AWS region and enter a bucket name of your choice. I suggest adding the suffix of your name or initials to your bucket name since each bucket name needs to be globally unique. Leave the rest as they are. Scroll down to the bottom and click on **Create bucket**.

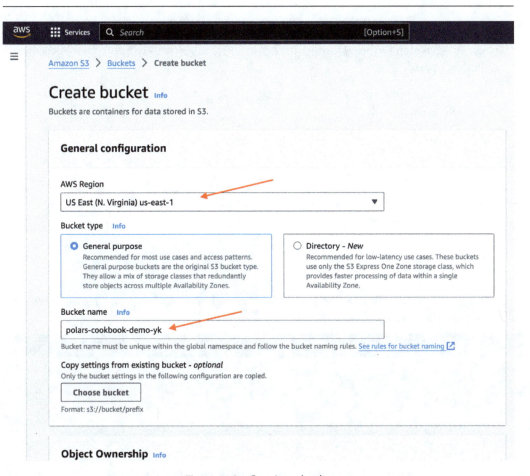

Figure 11.3 – Creating a bucket

5. Let's upload a file so we can test reading it from the Polars code. We'll upload the `titanic` dataset we used in *Chapter 1 Getting Started with Python Polars*. Navigate to your bucket and click on **Upload**.

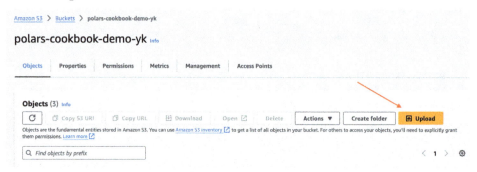

Figure 11.4 – Click on the Upload button

6. Drag and drop your file onto the screen. Scroll to the bottom and click on **Upload**. Leave the configurations as they are.

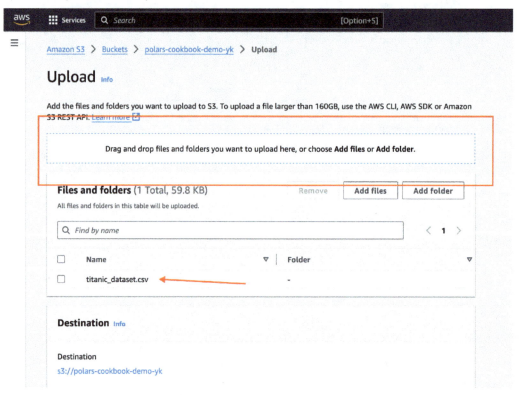

Figure 11.5 – Drag and drop the .csv file

7. Once you successfully upload your file, you'll see it in the list in your bucket.

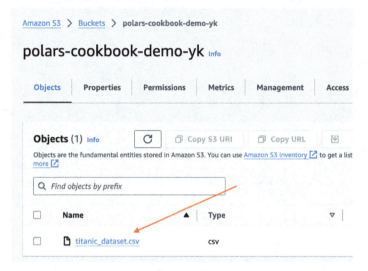

Figure 11.6 – A file has been added to the bucket

8. Let's prepare the necessary credentials to access S3 resources from your local machine. Head over to **IAM** and click on **Create user**:

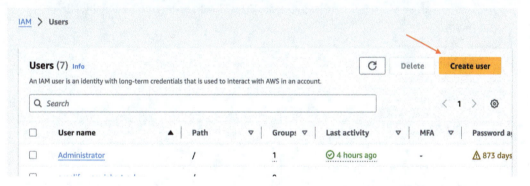

Figure 11.7 – Starting to create a new user

9. Enter your user name and click on **Next**:

Figure 11.8 – Entering user name

10. Choose **Attach policies directly**. Search for AmazonS3 in the search bar, and choose AmazonS3FullAccess https://docs.aws.amazon.com/aws-managed-policy/latest/reference/AmazonS3FullAccess.html). Finally, click on **Next**:

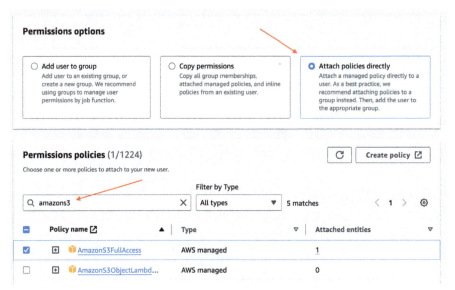

Figure 11.9 – Assigning an access policy for the new user

11. Click on **Create user** to finish creating a user with full access to S3:

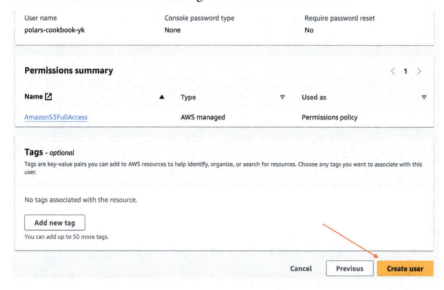

Figure 11.10 – Completing creating a new user

12. Now that you successfully created your user, go into your user's details, go to **Security credentials**, and click on **Create access key**:

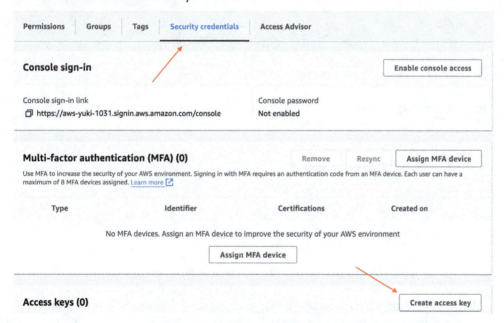

Figure 11.11 – Starting to create an access key

13. Choose **Local code**, check the box for confirmation, and click on **Next**:

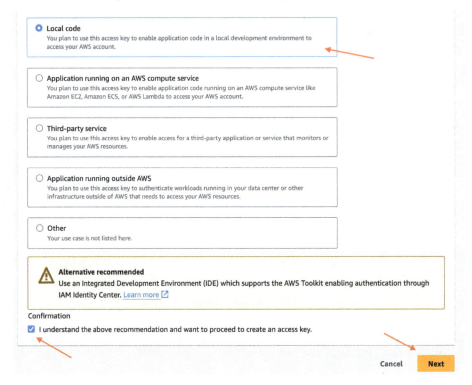

Figure 11.12 – Choosing the use case

14. Skip this phase and click on **Create access key**:

Figure 11.13 – Finishing up creating an access key

15. Make sure you copy your access key and secret access key for later use. Optionally, download the `.csv` file that contains those credentials:

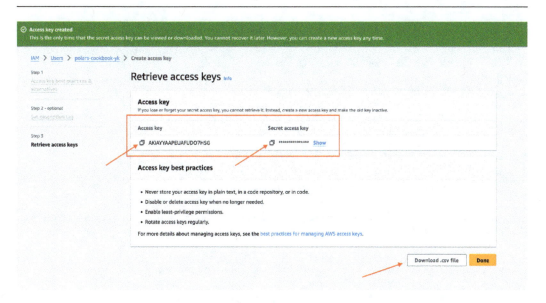

Figure 11.14 – Coping access key and secret access key

16. Head over to your IDE and make sure you install the S3FS library, which is a Python interface to S3. You also want to install PyArrow for one of the steps introduced in this recipe:

```
pip install s3fs pyarrow
```

> **Note**
>
> You will need to configure AWS CLI for *Step 1* and *Step 2* to work. Otherwise, please skip these steps or it'll throw an error. Configuring AWS CLI or any system is not as easy as we hope it is, and you will still understand the mechanics behind the code even if you skip *Step 1* and *Step 2*. With that being said, if you're still interested in setting up AWS CLI, please follow the steps on this documentation page: `https://docs.aws.amazon.com/cli/latest/userguide/cli-chap-getting-started.html`.
>
> After you set up AWS CLI, it stores your credentials in the `~/.aws/credentials` file. To learn more about ways to provide credentials, please refer to S3FS' documentation page: `https://s3fs.readthedocs.io/en/latest/#credentials`.

Now, you're ready to work with this file in your code using Polars!

How to do it...

Here's how to work with files in an S3 bucket in Polars:

1. Read a CSV file from an S3 bucket into a DataFrame:

    ```
    s3_file_path = 's3://polars-cookbook-demo-yk/titanic_dataset.
    csv'
    df = pl.read_csv(s3_file_path)
    df.head()
    ```

    The preceding code will return the following output:

    shape: (5, 12)

PassengerId	Survived	Pclass	Name	Sex	Age	SibSp	Parch	Ticket	Fare	Cabin	Embarked
i64	i64	i64	str	str	f64	i64	i64	str	f64	str	str
1	0	3	"Braund, Mr. Ow...	"male"	22.0	1	0	"A/5 21171"	7.25	null	"S"
2	1	1	"Cumings, Mrs. ...	"female"	38.0	1	0	"PC 17599"	71.2833	"C85"	"C"
3	1	3	"Heikkinen, Mis...	"female"	26.0	0	0	"STON/O2. 31012...	7.925	null	"S"
4	1	1	"Futrelle, Mrs....	"female"	35.0	1	0	"113803"	53.1	"C123"	"S"
5	0	3	"Allen, Mr. Wil...	"male"	35.0	0	0	"373450"	8.05	null	"S"

    Figure 11.15 – A snapshot of the titanic dataset

2. Write a DataFrame to a parquet file in an S3 bucket. For this, we need to utilize s3fs. Make sure you change the bucket path to what you configured:

    ```
    import s3fs

    fs = s3fs.S3FileSystem()
    s3_parquet_file_path = 's3://polars-cookbook-demo-yk/titanic_
    dataset.parquet'

    with fs.open(s3_parquet_file_path, mode='wb') as f:
        df.write_parquet(f)
    ```

    Go to your browser and check that the parquet file has been added:

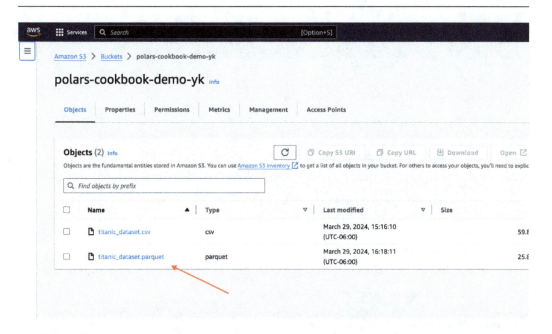

Figure 11.16 – The parquet file has been added to the bucket

3. Scan the parquet file into a LazyFrame. This time, we pass in the credentials manually:

```
storage_options= {
    'aws_access_key_id': 'YOUR_ACCESS_KEY_ID',
    'aws_secret_access_key': 'YOUR_SECRET_ACCESS_KEY',
    'aws_region': 'us-east-1'
}

lf = pl.scan_parquet(s3_parquet_file_path, storage_
options=storage_options)
lf.head().collect()
```

The preceding code will return the same output as *Figure 11.7*.

4. We will write to a parquet file again, but we will pass the credentials manually. As noted at the beginning of this chapter, it's not a good security practice to pass credentials directly in your code. This is just for demonstration purposes:

```
import s3fs

fs = s3fs.S3FileSystem(
    key='YOUR_ACCESS_KEY_ID',
    secret='YOUR_SECRET_ACCESS_KEY'
)
```

```
with fs.open(s3_parquet_file_path, mode='wb') as f:
    df.write_parquet(f)
```

5. You can also scan data with PyArrow:

```
import pyarrow.dataset as ds

dataset = ds.dataset(s3_parquet_file_path, format='parquet',
filesystem=fs)
df = (
    pl.scan_pyarrow_dataset(dataset)
    .filter(pl.col('Age') <= 30)
    .collect()
)
df.head()
```

The preceding code will return the following output:

shape: (5, 12)

PassengerId	Survived	Pclass	Name	Sex	Age	SibSp	Parch	Ticket	Fare	Cabin	Embarked
i64	i64	i64	str	str	f64	i64	i64	str	f64	str	str
1	0	3	"Braund, Mr. Ow...	"male"	22.0	1	0	"A/5 21171"	7.25	null	"S"
3	1	3	"Heikkinen, Mis...	"female"	26.0	0	0	"STON/O2. 31012...	7.925	null	"S"
8	0	3	"Palsson, Maste...	"male"	2.0	3	1	"349909"	21.075	null	"S"
9	1	3	"Johnson, Mrs. ...	"female"	27.0	0	2	"347742"	11.1333	null	"S"
10	1	2	"Nasser, Mrs. N...	"female"	14.0	1	0	"237736"	30.0708	null	"C"

Figure 11.17 – Filtered DataFrame read with PyArrow

How it works...

When you don't manually provide credentials in the `storage_options` parameter, Polars looks for the credentials file in the `.aws` directory or environmental variables. The `storage_options` parameter is what overrides the credentials if specified.

Polars can read files from S3 such as CSV, parquet, and **inter-process communication** (**IPC**). An advantage Polars has is that it can utilize its PyArrow integration. When Polars built-in functionality is not exactly what you need, you may be able to get it with PyArrow and then convert it to Polars.

When reading files in eager mode, Polars uses the `fsspec` library. Whereas, when you read files in lazy mode, Polars uses the `object_store` library in Rust.

See also

These are a few additional resources for learning more about reading from and writing to S3:

- `https://docs.pola.rs/user-guide/io/cloud-storage/`
- `https://s3fs.readthedocs.io/en/latest/`

Working with Azure Blob Storage

Azure has become popular as a choice of cloud computing platform in recent years. Azure Blob Storage is the object storage service on Azure.

In this recipe, we'll cover how to read from and write to Azure Blob Storage in Polars.

Getting ready

You need your Azure account for this recipe. Follow the instructions in the link I mentioned under *Technical requirements* to create your Azure account.

The following steps outline how to create an Azure Blob Storage account and upload a `.csv` file:

1. Login to your account at `https://azure.microsoft.com/`.

2. Navigate to creating a storage account.

Figure 11.18 – Azure home screen

3. Enter your subscription, resource group, and storage account name. Leave the rest as they are.

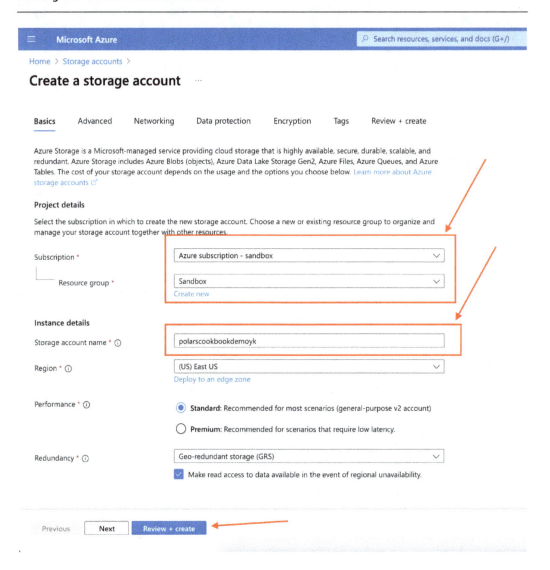

Figure 11.19 – Creating a storage account

4. You need to create a container to which we can upload our file. Click on **Container**, enter the name for it, and click on **Create**.

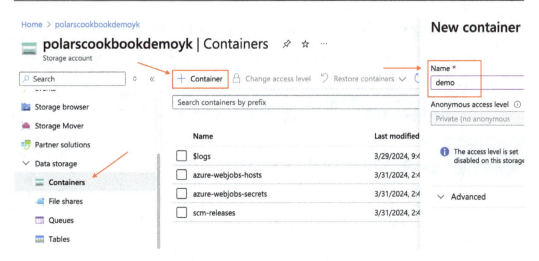

Figure 11.20 – Creating a container

5. Click into the demo container and upload the `titanic` dataset.

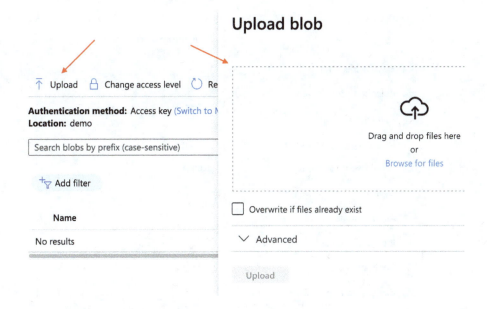

Figure 11.21 – Uploading a .csv file

We still need to establish a way to authenticate the client so you can have the read and write permissions to Azure Blob Storage. There are a few ways to accomplish that, but how we'll do that in this recipe is by using a service principle created by registering an application in Microsoft Entra ID (previously Azure Active Directory) on Azure and giving it the necessary permissions.

Here are the steps to register an app and give it permissions to Blob Storage:

1. Sign in to your Azure account. Go to Microsoft Entra ID and choose **App registrations**, which is found on the left pane.

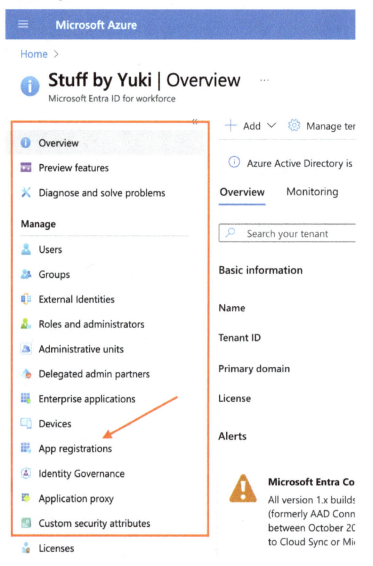

Figure 11.22 – Microsoft Entra ID home screen

2. Click on **New registration**.

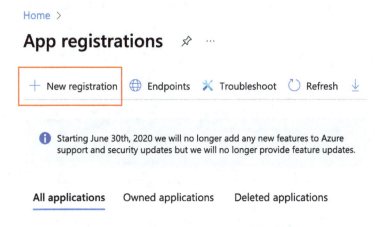

Figure 11.23 – Microsoft Entra ID App registration screen

3. Give it a name and leave the rest of the settings as they are. Click on **Register** to register an application.

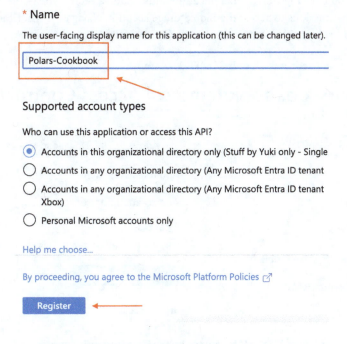

Figure 11.24 – The process of registering an application

4. After you register an application, we'll create the client credential. Go to **Certificates & secrets** and create a new client secret. Make sure you keep your secret ID and value for later use in your code:

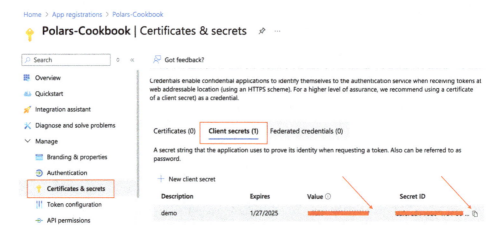

Figure 11.25 – The process of creating a client secret

5. Now that you have registered an application and created a secret, you're ready to give it the necessary permissions to access the Blob Storage account. Navigate to your Blob Storage and click on **Add role assignment** under **Access Control (IAM)**.

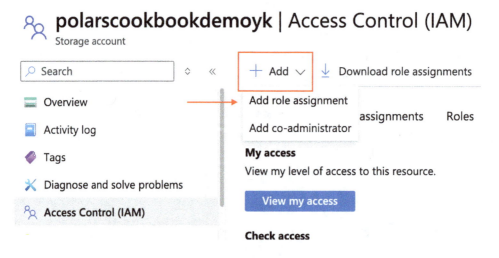

Figure 11.26 – IAM settings on the Blob Storage account

6. Type in `blob` in the search bar and choose **Storage Blob Data Contributor**, which gives you read, write, and delete access:

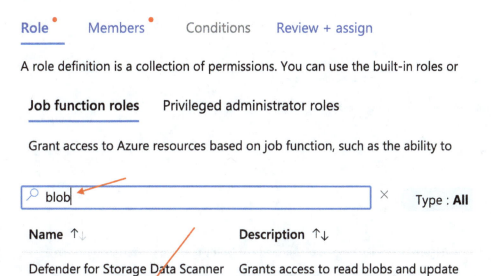

Figure 11.27 – The process of assigning a role

7. In the next screen, you'll select the app you just registered in Microsoft Entra ID. Then, you go to the next screen and click on **Review + assign** to complete the role assignment.

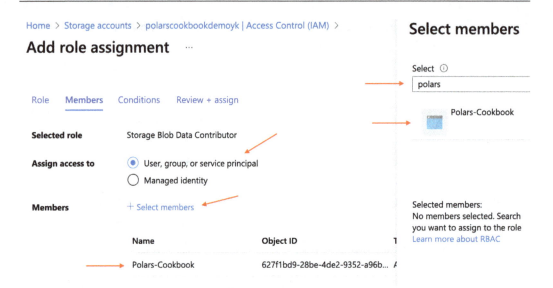

Figure 11.28 – The process of choosing an identity to which you assign the role to

You created your Azure Blob Storage account, uploaded a `.csv` file, and registered an application that has the necessary permissions to access Blob Storage.

The following is the list of credentials you'll be using in this recipe:

- `account_name`: The name of your storage account. Refer to *Figure 11.11*.

- `access_key`: The access key found in your storage account, under **Access keys**.

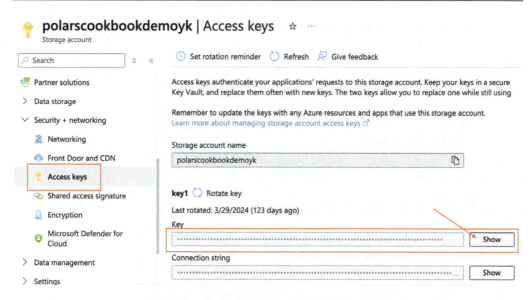

Figure 11.29 – Access key for the storage account

- `client_id`: Application (client) ID displayed in the overview page of your registered app.

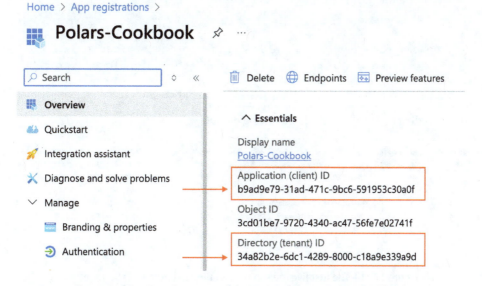

Figure 11.30 – Overview page of the registered app, Polars-Cookbook

- client_secret: The client secret value (**Value** not **Secret ID**) you generated in a previous step. Refer to *Figure 11.25*.

- tenant_id: Directory (tenant) ID shown on the overview page of your registered app. Refer to *Figure 11.29*. You can also get it on the overview page for the Microsoft Entra ID service.

8. Finally, let's make sure you install the ADLFS library:

```
pip install adlfs
```

How to do it...

Here's how to work with files in Azure Blob Storage:

1. Read a CSV file from Azure Blob Storage into a DataFrame. We'll pass in the credentials manually in the storage_options parameter:

```
storage_options={
    'account_name': 'YOUR_ACCOUNT_NAME',
    'access_key': 'YOUR_ACCOUNT_KEY',
    'client_id': 'YOUR_CLIENT_ID',
    'client_secret': 'YOUR_CLIENT_SECRET',
    'tenant_id': 'YOUR_TENANT_ID'
}

blob_csv_file_path = 'az://demo/titanic_dataset.csv'
df = pl.read_csv(blob_csv_file_path, storage_options=storage_options)
df.head()
```

The preceding code will return the following output:

shape: (5, 12)

PassengerId	Survived	Pclass	Name	Sex	Age	SibSp	Parch	Ticket	Fare	Cabin	Embarked
i64	i64	i64	str	str	f64	i64	i64	str	f64	str	str
1	0	3	"Braund, Mr. Ow...	"male"	22.0	1	0	"A/5 21171"	7.25	null	"S"
2	1	1	"Cumings, Mrs. ...	"female"	38.0	1	0	"PC 17599"	71.2833	"C85"	"C"
3	1	3	"Heikkinen, Mis...	"female"	26.0	0	0	"STON/O2. 31012...	7.925	null	"S"
4	1	1	"Futrelle, Mrs....	"female"	35.0	1	0	"113803"	53.1	"C123"	"S"
5	0	3	"Allen, Mr. Wil...	"male"	35.0	0	0	"373450"	8.05	null	"S"

Figure 11.31 – The first five rows in the .csv file read from Blob Storage

2. Write the DataFrame to Blob Storage:

```
import adlfs

fs = adlfs.AzureBlobFileSystem(
    account_name='YOUR_ACCOUNT_NAME',
    account_key='YOUR_ACCOUNT_KEY'
)


blob_parquet_file_path = 'az://demo/titanic_dataset.parquet'

with fs.open(blob_parquet_file_path, mode='wb') as f:
    df.write_parquet(f)
```

Let's check whether the parquet file has been added to Blob Storage:

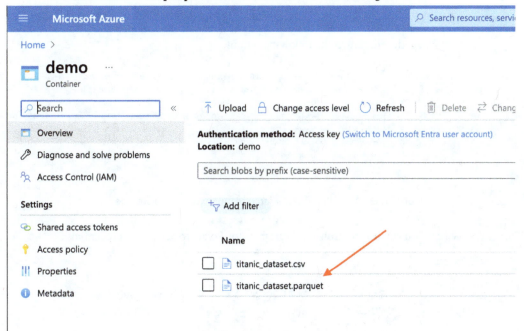

Figure 11.32 – Confirming the parquet file has been added

3. Scan a parquet file into a LazyFrame:

```
lf = pl.scan_parquet(blob_parquet_file_path, storage_
options=storage_options)
lf.head().collect()
```

The preceding code will return the same output as *Figure 11.20*.

4.   Scan a parquet file into a LazyFrame using PyArrow:

```
import pyarrow.dataset as ds

dataset = ds.dataset(
    blob_parquet_file_path,
    filesystem=fs,
    format='parquet')

df = (
    pl.scan_pyarrow_dataset(dataset)
    .filter(pl.col('Age') <= 30)
    .collect()
)

df.head()
```

The preceding code will return the following output:

shape: (5, 12)

PassengerId	Survived	Pclass	Name	Sex	Age	SibSp	Parch	Ticket	Fare	Cabin	Embarked
i64	i64	i64	str	str	f64	i64	i64	str	f64	str	str
1	0	3	"Braund, Mr. Ow...	"male"	22.0	1	0	"A/5 21171"	7.25	null	"S"
3	1	3	"Heikkinen, Mis...	"female"	26.0	0	0	"STON/O2. 31012...	7.925	null	"S"
8	0	3	"Palsson, Maste...	"male"	2.0	3	1	"349909"	21.075	null	"S"
9	1	3	"Johnson, Mrs. ...	"female"	27.0	0	2	"347742"	11.1333	null	"S"
10	1	2	"Nasser, Mrs. N...	"female"	14.0	1	0	"237736"	30.0708	null	"C"

Figure 11.33 – The filtered DataFrame scanned from Blob Storage

5.   Let's see how you can use environment variables for Polars to access the credentials:

```
import os

os.environ['AZURE_STORAGE_ACCOUNT_NAME'] = 'YOUR_ACCOUNT_NAME'
os.environ['AZURE_STORAGE_ACCOUNT_KEY'] = 'YOUR_ACCOUNT_KEY'

(
    pl.read_csv(blob_csv_file_path, storage_options=storage_
options)
    .head()
)
```

The preceding code will return the same output as *Figure 11.20*.

How it works...

The file path format you need to specify is `az://YOUR_CONTAINER_NAME/YOUR_FILE_NAME`. Just like when you worked with S3, the `storage_options` parameter is what you need to adjust to manually pass your credentials to access your Blob Storage account.

We showed an example of using a service principle to access an Azure resource, which, in our case, is Blob Storage. Make sure you choose the best way to authenticate your client when accessing Blob Storage.

When you work with Blob Storage in Polars using methods such as `.read_csv()` or `.scan_csv()`, `adlfs` is what's helping under the hood by passing a few parameters. It's a good idea to be familiar with `adlfs` to understand the available parameters and ways to work with Blob Storage.

There's more...

You can take the same approach to access Azure Data Lake Storage. I created an Azure Data Lake Storage account and gave the service principle the necessary permission as defined in *Getting ready*. The following code is how you read a `.csv` file from Azure Data Lake Storage. I just needed to modify the account name and account key:

```
storage_options['account_name'] = 'YOUR_ADLS_ACCOUNT_NAME'
storage_options['account_key'] = 'YOUR_ADLS_ACCOUNT_KEY'

df = pl.read_csv(blob_csv_file_path, storage_options=storage_options)
df.head()
```

The preceding code will return the same output as *Figure 11.20*.

See also

To learn more about working with Blob Storage in Polars, please refer to the following resources:

- `https://docs.pola.rs/user-guide/io/cloud-storage/`
- `https://pypi.org/project/adlfs/`

Working with Google Cloud Storage

Google Cloud Platform (GCP) is another major cloud provider in the industry. **Google Cloud Storage (GCS)** is an object file store on GCP. Polars can work with it by utilizing the GCSFS library.

In this recipe, we'll look at how to read from and write to a GCS in Polars.

Getting ready

You need your GCP account for this recipe. Follow the instructions in the link I mentioned under *Technical requirements* to create your GCP account.

Let's first create a Google Cloud Storage bucket and upload a file to it:

1.   Navigate to **Cloud Storage** and click on **CREATE**.

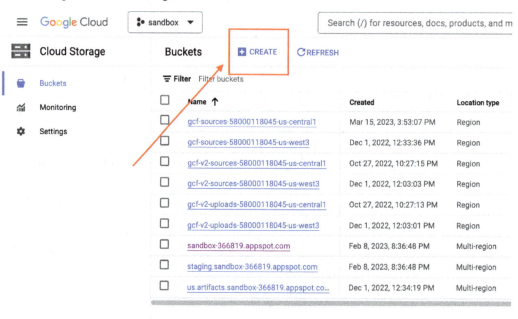

Figure 11.34 – Creating a cloud storage bucket

2. Enter your bucket name and leave the rest of the settings as they are. Note that each bucket name is globally unique. I suggest that you add your name or initials as a suffix. Once you have clicked **CREATE** to create a bucket, you may be asked to enforce public access prevention. Make sure to check this option to proceed.

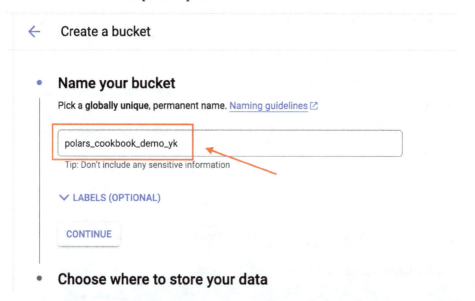

Figure 11.35 – Configuring bucket details

A popup will appear and make sure to keep the setting enabled and press **CONFIRM**:

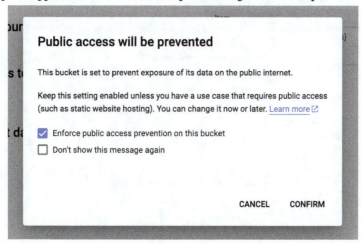

Figure 11.36 – The popup for enforcing public access prevention

3.   Once you created your cloud storage bucket, upload a `.csv` file.

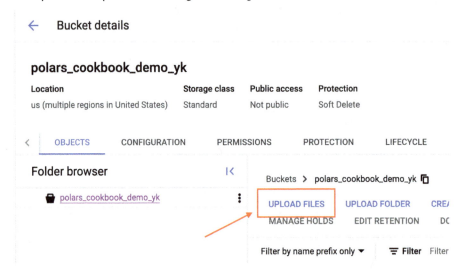

Figure 11.37 – Uploading a .csv file to a storage bucket

4.   Copy the URI of the file for later use.

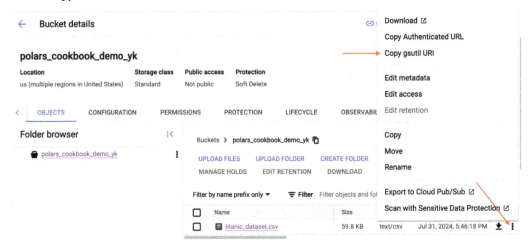

Figure 11.38 – Copying the file URI

There are several ways to authenticate the client, just like AWS and Azure. I've already configured a `gcloud` CLI on my machine, so Polars automatically looks for a JSON file that contains my credentials under `~/.config/gcloud` directory. You can set up a `gcloud` CLI by following this documentation: `https://cloud.google.com/sdk/docs/install/`. However, the `gcloud` CLI configuration is not needed as I'll demonstrate another way to authenticate and work with cloud storage, which is to use a service account key. To learn more about authentication, you can go to GCSFS's documentation page under **Credentials**: `https://gcsfs.readthedocs.io/en/latest/`.

You'd need a service account key for authentication for this recipe. You can follow this documentation to generate a JSON file containing a service account key along with other credentials: `https://cloud.google.com/sdk/docs/install/`. Here are the simplified steps to generate a JSON file with a service account key:

1. Go to the **IAM & Admin** page.

2. Click on **Service Accounts**.

3. Choose your service account and go to the **Keys** tab.

4. Click on **ADD KEY** and choose **Create new key**.

   Here is a screenshot of the previously described steps:

   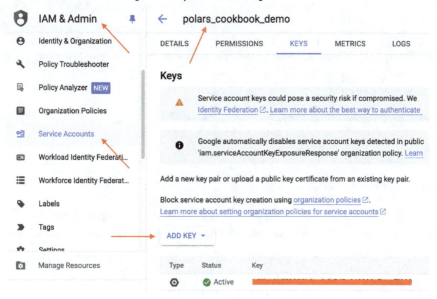

Figure 11.39 – Generating a service account key

Make sure your service account has appropriate permissions to read and write to your cloud storage bucket. You can go to your bucket and assign roles to your service account under the **PERMISSIONS** tab.

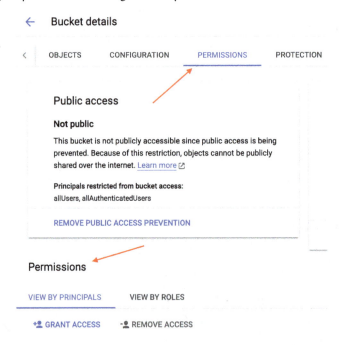

Figure 11.40 – Giving a service account permission to the cloud storage bucket

Finally, make sure you install `gcsfs` in the terminal:

```
pip install gcsfs
```

How to do it...

Here's how to read from and write to GCS:

1. Read a `.csv` file from a cloud storage bucket. Use the URI you copied for the file path. This code will look for the JSON file that contains your credentials as explained under *Getting ready*:

    ```
    gcs_csv_file_path = 'gs://polars_cookbook_demo_yk/titanic_
    dataset.csv'

    df = pl.read_csv(gcs_csv_file_path)
    df.head()
    ```

The preceding code will return the following output:

shape: (5, 12)

PassengerId	Survived	Pclass	Name	Sex	Age	SibSp	Parch	Ticket	Fare	Cabin	Embarked
i64	i64	i64	str	str	f64	i64	i64	str	f64	str	str
1	0	3	"Braund, Mr. Ow...	"male"	22.0	1	0	"A/5 21171"	7.25	null	"S"
2	1	1	"Cumings, Mrs. ...	"female"	38.0	1	0	"PC 17599"	71.2833	"C85"	"C"
3	1	3	"Heikkinen, Mis...	"female"	26.0	0	0	"STON/O2. 31012...	7.925	null	"S"
4	1	1	"Futrelle, Mrs....	"female"	35.0	1	0	"113803"	53.1	"C123"	"S"
5	0	3	"Allen, Mr. Wil...	"male"	35.0	0	0	"373450"	8.05	null	"S"

Figure 11.41 – The DataFrame read from the cloud storage bucket

2. Here's another way of passing the credentials when reading from cloud storage, which involves specifying the file path that contains your credentials:

```
credentials_file_path = 'YOUR_FILE_NAME.json'
storage_options = {'token': credentials_file_path}
df = pl.read_csv(gcs_csv_file_path, storage_options=storage_
options)
df.head()
```

The preceding code will return the same output as *Figure 11.30*.

3. Let's read the same file, but this time we'll decompose the contents in the JSON file and pass it as a Python dictionary:

```
import json

def read_json_into_dict(file_path):
    with open(file_path) as file:
        dict = json.load(file)
        return dict
storage_options = read_json_into_dict(credentials_file_path)

df = pl.read_csv(gcs_csv_file_path, storage_options=storage_
options)
df.head()
```

The preceding code will return the same output as *Figure 11.30*.

4. Write a parquet file to a cloud storage bucket. If you have a `gcloud` CLI configured, you may not need to specify the `token` parameter:

```
import gcsfs

fs = gcsfs.GCSFileSystem()

gcs_parquet_file_path = 'gs://polars_cookbook_demo_yk/titanic_
dataset.parquet'
```

```
with fs.open(gcs_parquet_file_path, mode='wb') as f:
    df.write_parquet(f)
```

5. Write a parquet file, passing credentials manually:

```
import gcsfs

fs = gcsfs.GCSFileSystem(
    token=credentials_file_path
)

with fs.open(gcs_parquet_file_path, mode='wb') as f:
    df.write_parquet(f)
```

6. Here's a way to read a file using PyArrow:

```
import pyarrow.dataset as ds

dataset = ds.dataset(
    gcs_parquet_file_path,
    filesystem=fs,
    format='parquet'
)

df = (
    pl.scan_pyarrow_dataset(dataset)
    .filter(pl.col('Age') <= 30)
    .collect()
)

df.head()
```

The preceding code will return the following output:

shape: (5, 12)

PassengerId	Survived	Pclass	Name	Sex	Age	SibSp	Parch	Ticket	Fare	Cabin	Embarked
i64	i64	i64	str	str	f64	i64	i64	str	f64	str	str
1	0	3	"Braund, Mr. Ow...	"male"	22.0	1	0	"A/5 21171"	7.25	null	"S"
3	1	3	"Heikkinen, Mis...	"female"	26.0	0	0	"STON/O2. 31012...	7.925	null	"S"
8	0	3	"Palsson, Maste...	"male"	2.0	3	1	"349909"	21.075	null	"S"
9	1	3	"Johnson, Mrs. ...	"female"	27.0	0	2	"347742"	11.1333	null	"S"
10	1	2	"Nasser, Mrs. N...	"female"	14.0	1	0	"237736"	30.0708	null	"C"

Figure 11.42 – The filtered DataFrame read from the cloud storage bucket

How it works...

When you have a `gcloud` CLI configured, it's easy as Polars looks for credentials automatically under the `~/.config/gcloud` directory. But the good news is that there are some other ways to authenticate the client to access your cloud storage bucket. One of those options is to pass a service account key to the `token` parameter. You can pass the file path or a dictionary containing key-value pairs of your credentials.

Just like the previous recipe, when you work with GCP in Polars using methods such as `.read_csv()` or `.scan_csv()`, `gcsfs` is what's helping under the hood. Understanding the available parameters and ways to work with a Google storage bucket with `gcsfs` will help you troubleshoot further when you get stuck.

> **Note**
>
> Just like when you work with other cloud providers, authentication can be tricky. Make sure to read the resources I listed in this recipe to better understand what you need to access your cloud storage from your Python code.

See also

To learn more about working with GCP in Polars, please refer to these resources:

- `https://docs.pola.rs/user-guide/io/cloud-storage/`
- `https://gcsfs.readthedocs.io/en/latest/`
- `https://cloud.google.com/iam/docs/keys-create-delete`
- `https://docs.rs/object_store/latest/object_store/gcp/enum.GoogleConfigKey.html`

Working with BigQuery

BigQuery is a popular option for a cloud data warehouse. It's part of GCP and its integrations with other Google technologies such as cloud functions, Google Analytics, and Looker Studio make your life easier.

When using pandas, you can utilize the `pandas-gbq` library and use methods such as `.read_gbq()` and `.to_gbq()` that help you work with BigQuery with ease. There are no such built-in methods in Polars, however, we'll try other approaches for how we can read from and write to BigQuery in this recipe.

Getting ready

You need to create a BigQuery dataset to work with. Here's how to create one:

1. Make sure to choose your project (in my case, it's `sandbox`) and go to BigQuery.

2. Click on the three dots and choose **Create dataset**.

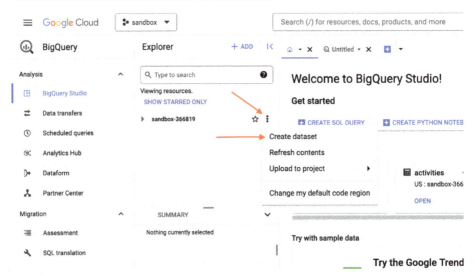

Figure 11.43 – Creating a new dataset

3. Enter your dataset name and click on **CREATE DATASET**.

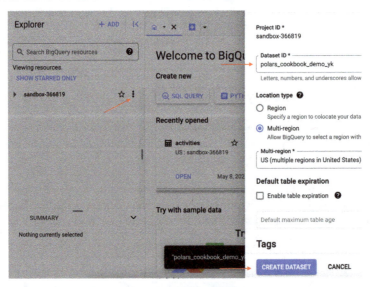

Figure 11.44 – Entering information about a new dataset

4. Once you have created a dataset, click on **+ ADD** and choose **Local file**. You can get data from GCP if you prefer that way.

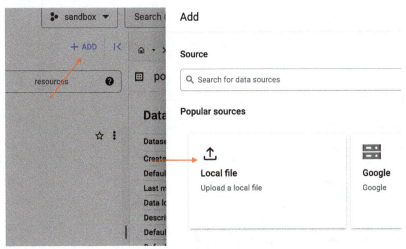

Figure 11.45 – Choosing to upload a file

5. Select your file in your local file system, enter your dataset and table, and check off the box for **Auto detect**. Finally, click on **CREATE TABLE**.

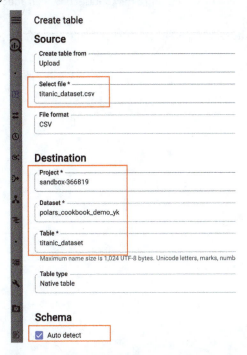

Figure 11.46 – Entering information about the file

6.   Confirm you have added a table under the dataset you just created.

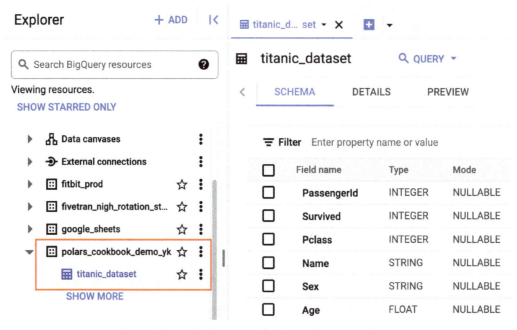

Figure 11.47 – Confirming a new dataset has been added

7.   Don't forget to copy the table ID we'll need to paste into our code.

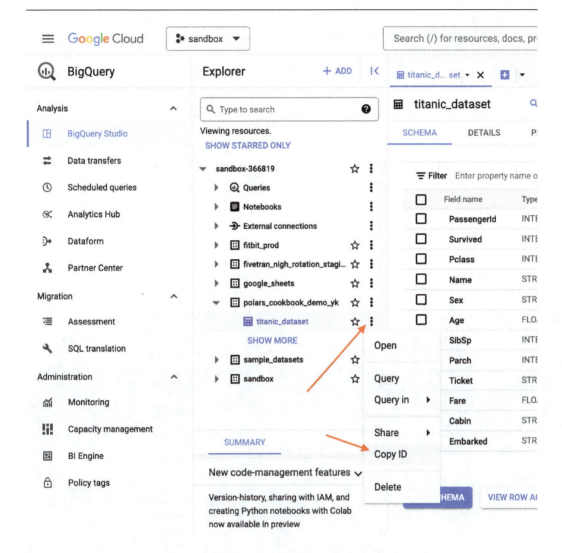

Figure 11.48 – Copying the dataset ID for later use

There are two approaches we'll be demonstrating in this recipe. One is using the `pl.read_database_uri()` function based on `connectorx`. Another approach is to use the BigQuery Python client library called `google-cloud-bigquery`.

For these two approaches, we'll need to install several libraries:

```
pip install pyarrow connectorx google-cloud-bigquery
```

With the approach of utilizing the BigQuery client library, you'll need to have `gcloud` set up for authentication. Be sure to follow this guide to set up `gcloud` for authentication: `https://googleapis.dev/python/google-api-core/latest/auth.html`

How to do it...

Here's how to work with BigQuery in Polars:

1. Read from a table in BigQuery using the `pl.read_database_uri()` function with `connectorx`:

```python
project = 'sandbox-366819'
dataset = 'polars_cookbook_demo_yk'
table = 'titanic_dataset'

query = f'''
    select *
    from {project}.{dataset}.{table}
'''

credentials_file_path = 'YOUR_CREDENTIALS_FILE_PATH'
uri = f'bigquery://{credentials_file_path}'

df = pl.read_database_uri(query, uri, engine='connectorx')
df.head()
```

The preceding code will return the following output:

shape: (5, 12)

PassengerId	Survived	Pclass	Name	Sex	Age	SibSp	Parch	Ticket	Fare	Cabin	Embarked
i64	i64	i64	str	str	f64	i64	i64	str	f64	str	str
180	0	3	"Leonard, Mr. L...	"male"	36.0	0	0	"LINE"	0.0	null	"S"
264	0	1	"Harrison, Mr. ...	"male"	40.0	0	0	"112059"	0.0	"B94"	"S"
278	0	2	"Parkes, Mr. Fr...	"male"	null	0	0	"239853"	0.0	null	"S"
303	0	3	"Johnson, Mr. W...	"male"	19.0	0	0	"LINE"	0.0	null	"S"
414	0	2	"Cunningham, Mr...	"male"	null	0	0	"239853"	0.0	null	"S"

Figure 11.49 – The first five rows of the titanic dataset read from BigQuery

2. Read from a table using the BigQuery client library:

```python
import polars as pl
from google.cloud import bigquery

client = bigquery.Client.from_service_account_json(credentials_
file_path)
query_job = client.query(query)
rows = query_job.result()

df = pl.from_arrow(rows.to_arrow())
df.head()
```

The preceding code will return the same output as *Figure 11.38*.

3. Write to a BigQuery table from a DataFrame using the BigQuery client library:

```python
import io

with io.BytesIO() as stream:
    df.write_csv(stream)
    stream.seek(0)
    job = client.load_table_from_file(
        stream,
        destination=f'{project}.{dataset}.titanic_dataset_v2',
        project=project,
        job_config=bigquery.LoadJobConfig(
            autodetect=True,
            source_format=bigquery.SourceFormat.CSV,
        ),
    )
job.result()
```

Let's confirm the dataset has been created:

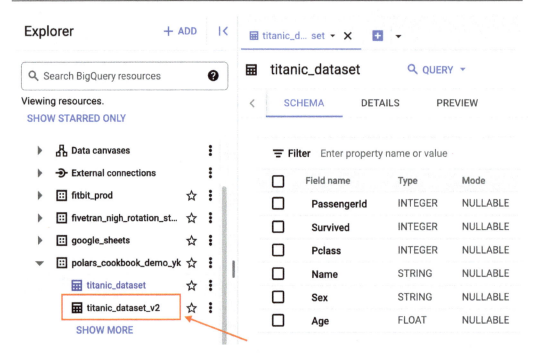

Figure 11.50 – Confirming the new dataset has been added

How it works...

All solutions explained in this recipe, `pl.read_database_uri()` with `connectorx`, `pl.read_database()`, and the BigQuery client library, are simple to implement. `connectorx` supports not only BigQuery, but also other databases such as Postgres, MySQL, Oracle, and SQL Server.

We chose a `.csv` file for the source format with `bigquery.SourceFormat.CSV`, but you can use other file types such as parquet and Avro that you can specify as in `bigquery.SourceFormat.PARQUET` and `bigquery.SourceFormat.AVRO`.

Also, we set the `autodetect` parameter to `True` in `bigquery.LoadJobConfig`. However, instead of doing that, you can manually specify your schema by manually entering your column names and data types.

If anything introduced in this recipe seems cumbersome, you can always decide to use methods such as the `.read_gbq()` and `.to_gbq()` methods with `pandas-gbq`.

> **Note**
>
> You could create your own authentication instead of using your authentication JSON file that contains your credentials. If you want to learn more about this approach, please refer to this Google Cloud documentation page: `https://cloud.google.com/docs/authentication/client-libraries#python`.

See also

To learn more about working with BigQuery in Polars, please refer to these resources:

- `https://sfu-db.github.io/connector-x/databases/bigquery.html`

- `https://cloud.google.com/docs/authentication/`

- `https://cloud.google.com/bigquery/docs/schema-detect`

- `https://cloud.google.com/bigquery/docs/schemas`

- `https://docs.pola.rs/user-guide/io/database/`

Working with Snowflake

Snowflake is another popular option for a cloud data warehouse. It's easy and quick to spin up a Snowflake instance and start working with it right away.

In this recipe, we'll be covering how to read from and write to tables in Snowflake.

Getting ready

You need a Snowflake account for this recipe. As of this writing, Snowflake gives you a 30-day free trial. Go ahead and create an account with this link: `https://signup.snowflake.com/`.

Let's create a database and a table we'll be reading data from:

1. Click on **Database** and enter your database name. Then, click on **Create**.

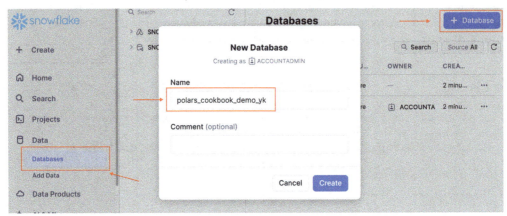

Figure 11.51 – Entering a database name

2. Click on **Schema** and enter your schema name. Click on **Create** to create a new schema.

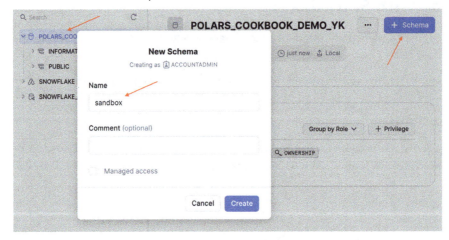

Figure 11.52 – Entering a schema name

3. Let's create a table. Choose your schema and click on **Create**. Choose **Table** from the dropdown. Then, select **From File**.

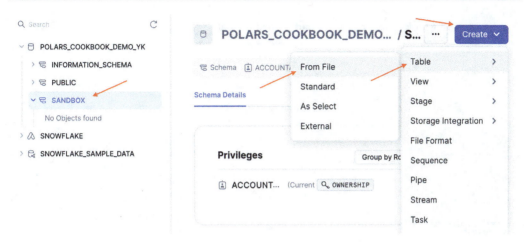

Figure 11.53 – Navigating to create a table

4. Choose your file to upload from your file system, enter your table name, and then click on **Next**.

Figure 11.54 – Uploading a file from a local machine

5.   If everything looks good, click on **Load** to load the data.

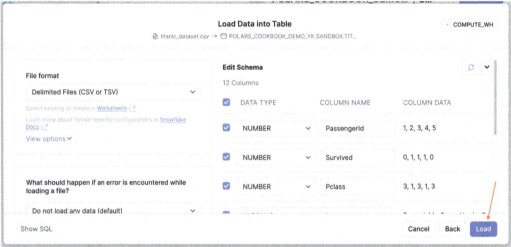

Figure 11.55 – Confirming the schema and starting the data load

A few approaches we'll look at in this recipe are as follows:

- Using **Arrow Database Connectivity (ADBC)**
- Using the Snowflake Python client library

You need the Snowflake ADBC driver and Snowflake Python client library installed:

```
pip install adbc-driver-snowflake snowflake-connector-python
```

How to do it...

Here's how to work with Snowflake:

1.   Read from a table in Snowflake using the `.read_database_uri()` method with the ADBC engine:

2. First, let's get the information you need to construct the URL. Here's how to get each piece of information:

- `username`: Your login username.

- `password`: Your login password.

- `account`: You can get your account identifier by going into your account. You simply hover over a few places to get to it. Your account identifier will look something like `NDJYDUA.TDB43776`. Make sure to replace the dot, `.`, with a hyphen, `-` to use it in your code. It should look like `NDJYDUA-TDB43776`.

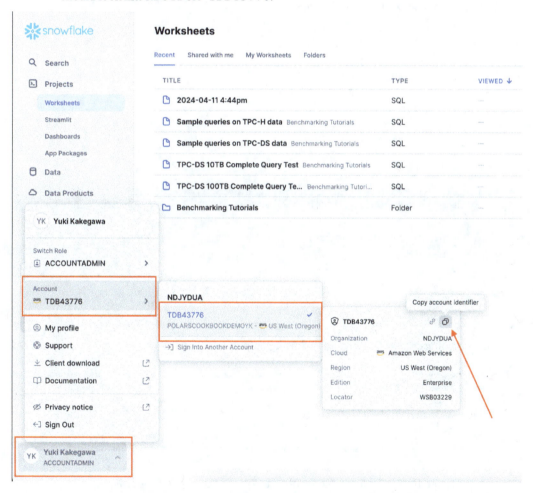

Figure 11.56 – A screen explaining how to get an account identifier

- **database**: We'll use POLARS_COOKBOOK_DEMO_YK that we just created.

- **warehouse**: We'll use COMPUTE_WH, which comes by default. You can create and choose a different warehouse if you'd like.

- **role**: We'll use ACCOUNTADMIN, which is pre-assigned to your default user. If you'd like to use some other role, you'll need to assign the role to a user in the admin console.

- **schema**: We'll use the SANDBOX schema that we just created.

- **table**: We'll use TITANIC_DATASET.

Here's the code snippet to read the data:

```
username = 'YOUR_USERNAME'
password = 'YOUR_PASS'
account = 'ndjydua-tdb43776'
database = 'POLARS_COOKBOOK_DEMO_YK'
warehouse = 'COMPUTE_WH'
role = 'ACCOUNTADMIN'
schema = 'SANDBOX'
table = 'TITANIC_DATASET'

query = f'select * from {table}'
uri = f'snowflake://{username}:{password}@{account}/{database}/
{schema}?warehouse={warehouse}&role={role}'

df = pl.read_database_uri(
    query,
    uri,
    engine='adbc'
)
df.head()
```

The preceding code will return the following output:

shape: (5, 12)

PASSENGERID	SURVIVED	PCLASS	NAME	SEX	AGE	SIBSP	PARCH	TICKET	FARE	CABIN	EMBARKED
f64	f64	f64	str	str	f64	f64	f64	str	f64	str	str
1.0	0.0	3.0	"Braund, Mr. Ow...	"male"	22.0	1.0	0.0	"A/5 21171"	7.25	null	"S"
2.0	1.0	1.0	"Cumings, Mrs. ...	"female"	38.0	1.0	0.0	"PC 17599"	71.2833	"C85"	"C"
3.0	1.0	3.0	"Heikkinen, Mis...	"female"	26.0	0.0	0.0	"STON/O2. 31012...	7.925	null	"S"
4.0	1.0	1.0	"Futrelle, Mrs....	"female"	35.0	1.0	0.0	"113803"	53.1	"C123"	"S"
5.0	0.0	3.0	"Allen, Mr. Wil...	"male"	35.0	0.0	0.0	"373450"	8.05	null	"S"

Figure 11.57 – The first five rows from the DataFrame read from Snowflake

3. Read data into a DataFrame using the Snowflake Python client library:

```
import snowflake.connector

conn = snowflake.connector.connect(
    user=username,
    password=password,
    account=account,
    warehouse=warehouse,
    database=database,
    schema=schema
)


(
    pl.from_arrow(conn.cursor().execute(query).fetch_arrow_
all())
    .head()
)
```

The preceding code will return the same output as *Figure 11.46*.

4. Read data using the `.read_database()` method by passing the Snowflake connection we just created in *Step 2*:

```
df = pl.read_database(query, connection=conn)
df.head()
```

The preceding code will return the same output as *Figure 11.46*.

How it works...

There are multiple ways to construct the URI to connect to Snowflake. You can learn more about the URI format in this ADBC documentation: `https://arrow.apache.org/adbc/0.5.1/driver/snowflake.html`

When using the Snowflake Python Client library, you can fetch your query result into formats such as Arrow, pandas, and a list of dictionaries. In our example, we fetched the result into Arrow as Polars allows us to utilize zero-copy with Arrow.

In *Step 3*, we showed a good way to use the `.read_database()` method, using the existing connection instead of constructing the URI on your own.

See also

To learn more about working with Snowflake in Polars, please refer to these resources:

- `https://docs.pola.rs/user-guide/io/database/`

- `https://arrow.apache.org/adbc/0.5.1/driver/snowflake.html`

- `https://docs.pola.rs/py-polars/html/reference/api/polars.read_database_uri.html`

- `https://docs.pola.rs/api/python/stable/reference/api/polars.read_database.html`

- `https://docs.snowflake.com/en/developer-guide/python-connector/python-connector-api`

12

Testing and Debugging in Polars

We've covered many aspects of data processing and manipulations in Python Polars up so far. We'll continue learning about what Polars can do in this chapter. The concepts taught in this chapter may apply to any data workflow, as testing and debugging are crucial components that help ensure the quality of your code.

This chapter contains the following recipes:

- Debugging chained operations
- Inspecting and optimizing the query plan
- Testing data quality with cuallee
- Running unit tests with pytest

Technical requirements

You can download the datasets and code in the GitHub repository.

- Data: `https://github.com/PacktPublishing/Polars-Cookbook/tree/main/data`
- Code: `https://github.com/PacktPublishing/Polars-Cookbook/tree/main/Chapter12`

It is assumed that you have installed the Polars library in your Python environment:

```
>>> pip install polars
```

It is also assumed that you have imported it into your code:

```
import polars as pl
```

Read the Pokemon dataset in a LazyFrame that will be used throughout this chapter:

```
lf = pl.scan_csv('../data/pokemon.csv')
```

Debugging chained operations

Throughout this book, we've used method chaining, whereby multiple operations are stacked one after another instead of assigning each operation to a variable each time. Method chaining not only removes the repetitive code but also helps you design your code with relevant operations in a chain. This helps readers understand your code easily and reduces the mental burden of making sense of the context of your logic.

The potential downsides of method chaining are readability and difficulty in debugging. The former can be solved by not chaining operations horizontally. When you put each operation on a new line, readability will be improved. What about debugging? The same thing is applied there. When you separate each operation on a new line, you'll be able to inspect exactly why an error is occurring, especially with tools such as debugger in your IDE.

In this recipe, we'll cover how you can troubleshoot, test, and debug your chained operations in Polars.

How to do it...

Here's how to debug chained operations.

1. The easiest and simplest thing you can do when debugging your code is to comment out a portion of your chain. Let's apply some operations in a chain as an example:

```
(
    lf
    .with_columns(
        pl.col('Attack').rank(method='dense').alias('Atk Rank'),
        pl.col('Defense').rank(method='dense').alias('Def
Rank'),
        pl.col('Speed').rank(method='dense').alias('Spe Rank'),
    )
    .select(
        'Name',
        'Total',
        'Attack',
        'Defense',
```

```
            'Speed',
            pl.col('^*Rank$')
        )
    .sort('Total')
    .head()
    .collect()
)
```

The preceding code will return the following output:

shape: (5, 8)

Name	Total	Attack	Defense	Speed	Atk Rank	Def Rank	Spe Rank
str	i64	i64	i64	i64	u32	u32	u32
"Caterpie"	195	30	35	45	5	6	10
"Weedle"	195	35	30	50	6	5	12
"Magikarp"	200	10	55	80	2	16	27
"Kakuna"	205	25	50	35	4	13	5
"Metapod"	205	20	55	30	3	16	4

Figure 12.1 – The first five rows of the query

2. Now, let's modify the query so that it raises an error, which we'll troubleshoot:

```
(
    lf
    .with_columns(
        pl.col('Attack').rank(method='dense').alias('Atk Rank'),
        pl.col('Defense').rank(method='dense').alias('Def
Rank'),
        pl.col('Speed').rank(method='dense').alias('Spe Rank'),
    )
    .select(
        'Name',
        'Total',
        'Attack',
        'Deffense',
        'Speed',
        pl.col('^*Rank$')
    )
    .sort('Total')
    .head()
```

```
    .collect()
)
```

The preceding code will return the following output:

```
>>
ColumnNotFoundError: Deffense
This error occurred with the following context stack:
[1] 'select' failed
[2] 'sort' input failed to resolve
[3] 'slice' input failed to resolve
```

3. You can see that in the preceding step, there is an error somewhere in your query. If you look at the error message, it's pretty clear that the issue exists after the `.with_columns()` method and it's the misspelling of the word `Defense`. However, for the sake of demonstration, let's try commenting out a portion of your code to manually test to identify what could be wrong:

```
(
    lf
    .with_columns(
        pl.col('Attack').rank(method='dense').alias('Atk Rank'),
        pl.col('Defense').rank(method='dense').alias('Def
Rank'),
        pl.col('Speed').rank(method='dense').alias('Spe Rank'),
    )
    # .select(
    #     'Name',
    #     'Total',
    #     'Attack',
    #     'Deffense',
    #     'Speed',
    #     pl.col('^*Rank$')
    # )
    .sort('Total')
    .head()
    .collect()
)
```

The preceding code will return the following output:

shape: (5, 16)

#	Name	Type 1	Type 2	Total	HP	Attack	Defense	Sp. Atk	Sp. Def	Speed	Generation	Legendary	Atk Rank	Def Rank	Spe Rank
i64	str	str	str	i64	i64	i64	i64	i64	i64	i64	i64	bool	u32	u32	u32
10	"Caterpie"	"Bug"	null	195	45	30	35	20	20	45	1	false	5	6	10
13	"Weedle"	"Bug"	"Poison"	195	40	35	30	20	20	50	1	false	6	5	12
129	"Magikarp"	"Water"	null	200	20	10	55	15	20	80	1	false	2	16	27
14	"Kakuna"	"Bug"	"Poison"	205	45	25	50	25	25	35	1	false	4	13	5
11	"Metapod"	"Bug"	null	205	50	20	55	25	25	30	1	false	3	16	4

Figure 12.2 – The DataFrame with new columns

Okay. I have just commented out a portion that I suspected was the issue. Now my code ran successfully, so I can say that I have confirmed where the issue exists.

That's an example of how you can manually comment or uncomment blocks of your code to identify issues in your code.

Let's see how we can use debugger in VSCode this time. Debugger doesn't work well with a LazyFrame since everything gets executed when .collect() is called. So, we'll use a DataFrame in this step.

1. Let's add debug points in two places, one on the .with_columns() method and another on the .select() method:

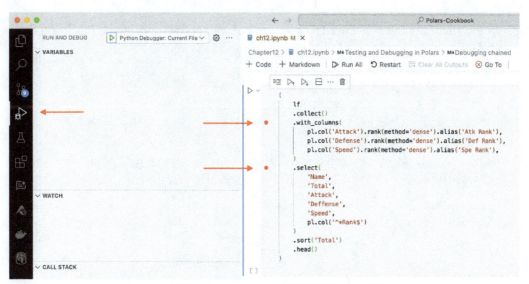

Figure 12.3 – The query to which debug points are added

2. After you finish going over the `select` operation of your code in debugger, it will give you an error right away:

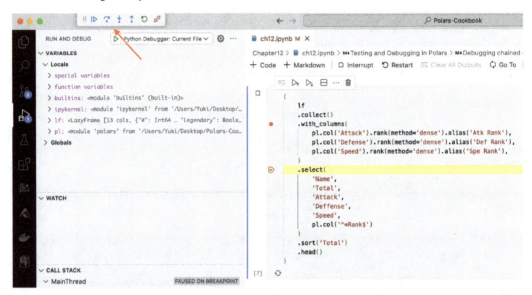

Figure 12.4 – The select operation right at the end of debugging

3. Right after stepping over the select operation of your code in a debugger, you'll get the following error:

```
>>
ColumnNotFoundError: Deffense
Error originated just after this operation:
DF ["#", "Name", "Type 1", "Type 2"]; PROJECT */16 COLUMNS;
SELECTION: "None"
```

If you run the same in a Python script, the error message pops up right on the spot where it happened:

```
1      import polars as pl
2
3      lf = pl.scan_csv('data/pokemon.csv')
4
5      (
6          lf
7          .collect()
●   8          .with_columns(
9              pl.col('Attack').rank(method='dense').alias('Atk Rank'),
10             pl.col('Defense').rank(method='dense').alias('Def Rank'),
11             pl.col('Speed').rank(method='dense').alias('Spe Rank'),
12         )
▷  13         . ▷ select(
```

```
Exception has occurred: ColumnNotFoundError  ✕
Deffense

Error originated just after this operation:
DF ["#", "Name", "Type 1", "Type 2"]; PROJECT */16 COLUMNS; SELECTION: "None"

  File "/Users/Yuki/Desktop/Polars-Cookbook/Chapter12/playground.py", line 13, in <module>
    .select(
    ^^^^^^^
polars.exceptions.ColumnNotFoundError: Deffense

Error originated just after this operation:
DF ["#", "Name", "Type 1", "Type 2"]; PROJECT */16 COLUMNS; SELECTION: "None"
```

```
14             'Name',
15             'Total',
16             'Attack',
17             'Deffense',
18             'Speed',
19             pl.col('^*Rank$')
20         )
21         .sort('Total')
22         .head()
23     )
24
```

Figure 12.5 – An error message right on the place where it happened

How it works...

Testing and debugging your code by commenting line by line is probably the easiest and most convenient option in your development, especially when you're still exploring your data.

Using a DataFrame is a better fit for a debugger, as you will get an error exactly where it occurs. It doesn't make as much sense to use a debugger on a LazyFrame alone, as everything comes down to the .collect() method call. However, the error message is often detailed enough to figure out where the issue lies, and you may not need a debugger in the first place.

There's more...

I like using the `.pipe()` method to group similar operations together in a function. Although this doesn't help you debug more easily than the approach you just saw in this recipe, it makes the code more readable and maintainable. It also separates each logic into organized blocks.

Here's an example of how you can use the `.pipe()` method in your code:

```
def add_ranks(lf: pl.LazyFrame) -> pl.LazyFrame:
    return (
        lf
        .with_columns(
            pl.col('Attack').rank(method='dense').alias('Atk Rank'),
            pl.col('Defense').rank(method='dense').alias('Def Rank'),
            pl.col('Speed').rank(method='dense').alias('Spe Rank'),
        )
    )


def keep_cols(lf: pl.LazyFrame) -> pl.LazyFrame:
    return (
        lf
        .select(
            'Name',
            'Total',
            'Attack',
            'Defense',
            'Speed',
            pl.col('^*Rank$')
        )
    )


(
    lf
    .pipe(add_ranks)
    .pipe(keep_cols)
    .sort('Total')
    .head()
    .collect()
)
```

The preceding code will return the same output as in *Figure 12.2*.

See also

You can refer to these resources for learning more about method chaining:

- `https://quanticdev.com/articles/method-chaining/`
- `https://kevinheavey.github.io/modern-polars/method_chaining.html`
- `https://ponder.io/professional-pandas-the-pandas-assign-method-and-chaining/`
- `https://betterprogramming.pub/an-interview-with-python-and-pandas-trainer-matt-harrison-e000374ddab`

Inspecting and optimizing the query plan

One feature that Polars provides that tools such as pandas don't is the ability to inspect the query plan. Lazy evaluation in LazyFrames makes it possible for Polars to optimize the query plan and for us to inspect what it's trying to do in a query.

In this recipe, we'll look at how we can inspect and optimize the query plan.

Getting ready

You need the `graphviz` library installed on your machine to use one of the methods for inspecting the query plan, which is the `.show_graph()` method. However, the installation process is a bit different than other Python libraries because you need to add a path to `graphviz`.

On Mac, the easiest way to do all that at once is using Homebrew, a package manager. You just need to run the following and the installation will be complete:

```
brew install graphviz
```

If you have a Windows machine or want to know more about the installation process, please refer to this documentation page: `https://pygraphviz.github.io/documentation/stable/install.html`.

If you just install the library in `pip` and didn't add the path to `graphviz`, then you'll get the following error:

```
ImportError: Graphviz dot binary should be on your PATH
```

Let's add a function for additional transformations:

```
def keep_grass_or_fire(lf):
    accepted_types = ['Grass', 'Fire']
    return (
        lf
        .filter(
            (pl.col('Type 1').is_in(accepted_types))
            | (pl.col('Type 2').is_in(accepted_types))
        )
    )
```

Also, make sure to install the `matplotlib` library for `.show_graph()` to work:

```
pip install matplotlib
```

How to do it...

Here's how to inspect and optimize the query plan.

1. Use the `.show_graph()` method to visualize the query plan:

    ```
    (
        lf
        .pipe(add_ranks)
        .pipe(keep_grass_or_fire)
        .pipe(keep_cols)
        .show_graph()
    )
    ```

    The preceding code will return the following output:

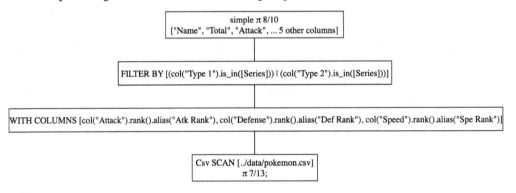

Figure 12.6 – The visualized query plan with the show_graph method

Notice that in the very first operation, which is scanning the `.csv` file, you're only reading in the necessary columns (7 out of 13 columns). This is due to an optimization method called **projection pushdown**, which pushes the column selection as upstream as possible.

The pi symbol indicates the column selections, and the sigma symbol indicates the filter conditions.

2. Set the optimized parameter to `False` to see the non-optimized query plan:

```
(
    lf
    .pipe(add_ranks)
    .pipe(keep_grass_or_fire)
    .pipe(keep_cols)
    .show_graph(optimized=False)
)
```

The preceding code will return the following output:

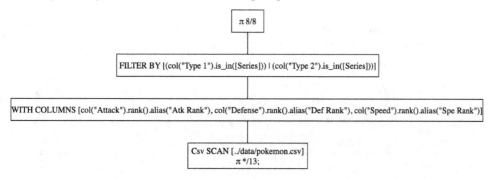

Figure 12.7 – The visualized non-optimized query plan with the show_graph method

Notice that you're reading all the columns at the scan operation, whereas in the preceding step, you only read seven columns.

3. Use the `.explain()` method to print out the query plan in a text format. You need to use the Python `print()` function to format the output text:

```
print(
    lf
    .pipe(add_ranks)
    .pipe(keep_grass_or_fire)
    .pipe(keep_cols)
    .explain()
)
```

The preceding code will return the following output:

```
>>
 simple π 8/10 ["Name", "Total", "Attack", ... 5 other columns]
  FILTER [(col("Type 1").is_in([Series])) | (col("Type 2").is_
in([Series]))] FROM
     WITH_COLUMNS:
     [col("Attack").rank().alias("Atk Rank"), col("Defense").
rank().alias("Def Rank"), col("Speed").rank().alias("Spe Rank")]
        Csv SCAN [../data/pokemon.csv]
        PROJECT 7/13 COLUMNS
```

4. The `.explain()` method also allows you to turn certain optimizations on or off. Let's see the non-optimized query plan:

```
print(
    lf
    .pipe(add_ranks)
    .pipe(keep_grass_or_fire)
    .pipe(keep_cols)
    .explain(optimized=False)
)
```

The preceding code will return the following output:

```
>>
 SELECT [col("Name"), col("Total"), col("Attack"),
col("Defense"), col("Speed"), col("Atk Rank"), col("Def Rank"),
col("Spe Rank")] FROM
  FILTER [(col("Type 1").is_in([Series])) | (col("Type 2").is_
in([Series]))] FROM
     WITH_COLUMNS:
     [col("Attack").rank().alias("Atk Rank"), col("Defense").
rank().alias("Def Rank"), col("Speed").rank().alias("Spe Rank")]
        Csv SCAN [../data/pokemon.csv]
        PROJECT */13 COLUMNS
```

5. You might've noticed that we have a filter applied in the query (the `keep_grass_or_fire` function), but it is not pushed to the scan operation. This can happen even though Polars does its best to optimize without user interventions. Let's switch up the order of transformations to see how we might optimize the query plan. We'll apply filtering conditions and column selections first:

```
(
    lf
    .pipe(keep_grass_or_fire)
    .pipe(keep_cols)
```

```
    .pipe(add_ranks)
    .show_graph()
)
```

The preceding code will return the following output:

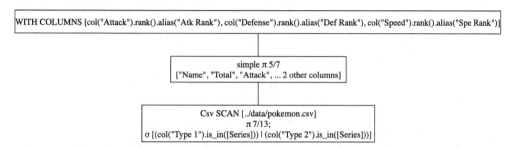

Figure 12.8 – The optimized plan with both predicate pushdown and projection pushdown

We have successfully optimized the query for the Polars engine to recognize the filtering conditions!

How it works...

Both `.show_graph()` and `.explain()` show the optimized query plan by default. There are parameters such as `optimized`, `predicate_pushdown`, `projection_pushdown`, `slice_pushdown`, and `comm_subplan_elim` available. You can set these to *False* to turn off certain optimizations from the query plan.

There's more...

One parameter that's worth demonstrating is the `streaming` parameter. It lets you know which part of your query is run in streaming mode.

Here's an example:

```
print(
    lf
    .pipe(keep_grass_or_fire)
    .pipe(keep_cols)
    .pipe(add_ranks)
    .explain(streaming=True)
)
```

The preceding code will return the following output:

```
>>
WITH_COLUMNS:
[col("Attack").rank().alias("Atk Rank"), col("Defense").rank().
alias("Def Rank"), col("Speed").rank().alias("Spe Rank")]
  STREAMING:
    simple π 5/7 ["Name", "Total", "Attack", ... 2 other columns]
      Csv SCAN [data/pokemon.csv]
      PROJECT 7/13 COLUMNS
      SELECTION: [(col("Type 1").is_in([Series])) | (col("Type 2").
is_in([Series]))]
```

Everything below the STREAMING keyword is run in streaming mode. In our query, those rank operations are *not* run in streaming mode. As you can see, this is helpful in identifying the bottlenecks for running the parts of the query in a streaming fashion.

See also

These are additional resources for inspecting and optimizing the query plan:

- https://pygraphviz.github.io/documentation/stable/install.html
- https://docs.pola.rs/user-guide/lazy/query-plan/
- https://docs.pola.rs/user-guide/lazy/optimizations/
- https://docs.pola.rs/user-guide/lazy/execution/
- https://docs.pola.rs/py-polars/html/reference/lazyframe/api/polars.LazyFrame.explain.html
- https://docs.pola.rs/py-polars/html/reference/lazyframe/api/polars.LazyFrame.show_graph.html

Testing data quality with cuallee

There are many data quality and validation tools out there for data workflows in Python. One of the most popular options is **Great Expectations**. There is another popular solution called deequ. cuallee was born to improve deequ by making the API more user-friendly and cost-efficient. It is a very lightweight library that helps you conduct data quality checks quickly in your workflows. cuallee's API is clean and easy to understand, so you can get started with data quality testing using cuallee without facing many technical hurdles.

In this recipe, we'll cover how we can add data quality tests to your data transformation process in Polars.

Getting ready

Install `cuallee` in your terminal with the following code:

```
pip install cuallee
```

How to do it...

Here's how to conduct data quality checks with `cuallee`.

1. `cuallee` only takes a DataFrame as its input, so let's convert our LazyFrame to a DataFrame:

    ```
    df = lf.collect()
    df.head()
    ```

    The preceding code will return the following output:

    shape: (5, 13)

#	Name	Type 1	Type 2	Total	HP	...	Defense	Sp. Atk	Sp. Def	Speed	Generation	Legendary
i64	str	str	str	i64	i64	...	i64	i64	i64	i64	i64	bool
1	"Bulbasaur"	"Grass"	"Poison"	318	45	...	49	65	65	45	1	false
2	"Ivysaur"	"Grass"	"Poison"	405	60	...	63	80	80	60	1	false
3	"Venusaur"	"Grass"	"Poison"	525	80	...	83	100	100	80	1	false
3	"VenusaurMega V..."	"Grass"	"Poison"	625	80	...	123	122	120	80	1	false
4	"Charmander"	"Fire"	null	309	39	...	43	60	50	65	1	false

    Figure 12.9 – The first five rows in the Pokemon dataset

2. Check for completeness and uniqueness on the `Name` column:

    ```
    from cuallee import Check, CheckLevel
    check = Check(CheckLevel.WARNING, 'Completeness')
    (
        check
        .is_complete('Name')
        .is_unique('Name')
        .validate(df)
    )
    ```

    The preceding code will return the following output:

    shape: (2, 12)

id	timestamp	check	level	column	rule	value	rows	violations	pass_rate	pass_threshold	status
i64	str	str	str	str	str	str	i64	i64	f64	f64	str
1	"2024-03-21 22:..."	"Completeness"	"WARNING"	"('Type 1', 'Ty..."	"is_complete"	"N/A"	163	0	1.0	1.0	"PASS"
2	"2024-03-21 22:..."	"Completeness"	"WARNING"	"Name"	"is_unique"	"N/A"	163	0	1.0	1.0	"PASS"

    Figure 12.10 – The output DataFrame with validation results

    How nice is it that `cuallee` returns a DataFrame that contains the validation result!

3. Check the accepted values in a column and only return the columns we need from the output DataFrame:

```
check = Check(CheckLevel.WARNING, 'Accepted Values')
accepted_types = (
    lf.select('Type 1')
    .unique()
    .collect()
    .to_series()
    .to_list()
)


(
    check
    .is_contained_in('Type 1', set(accepted_types))
    .validate(df)
    .select('check', 'column', 'rule', 'status')
)
```

The preceding code will return the following output:

```
shape: (1, 4)
```

check	column	rule	status
str	str	str	str
"Accepted Value...	"Type 1"	"is_contained_i...	"PASS"

Figure 12.11 – The output DataFrame with selected columns

4. To test on a set or group of columns, you just need to pass in your columns as a list:

```
check = Check(CheckLevel.WARNING, 'Validation on Stats')
stats_cols = [
    'HP',
    'Attack',
    'Defense',
    'Sp. Atk',
    'Sp. Def',
    'Speed'
]
res_cols = [
    'check',
    'column',
    'rule',
    'rows',
```

```
        'violations',
        'pass_rate',
        'status'
    ]

    (
        check
        .is_complete(stats_cols)
        .is_greater_than(stats_cols, 0)
        .validate(df)
        .select(res_cols)
    )
```

The preceding code will return the following output:

shape: (2, 7)

check	column	rule	rows	violations	pass_rate	status
str	str	str	i64	i64	f64	str
"Validation on ...	"('HP', 'Attack...	"is_complete"	163	0	1.0	"PASS"
"Validation on ...	"('HP', 'Attack...	"is_greater_tha...	163	0	1.0	"PASS"

Figure 12.12 – The output DataFrame validations on stats

5. You can test for completeness and uniqueness in multiple columns at once:

```
check = Check(CheckLevel.WARNING, 'Completeness')
cols = ['Name', 'Type 1', 'Type 2']
(
    check
    .are_complete(cols)
    .are_unique(cols)
    .validate(df)
    .select(
        'check',
        'column',
        'rule',
        'rows',
```

```
        'violations',
        'pass_rate',
        'status'
    )
)
```

The preceding code will return the following output:

shape: (2, 7)

check	column	rule	rows	violations	pass_rate	status
str	str	str	i64	f64	f64	str
"Completeness"	"('Name', 'Type...	"are_complete"	163	28.666667	0.824131	"FAIL"
"Completeness"	"('Name', 'Type...	"are_unique"	163	107.666667	0.339468	"FAIL"

Figure 12.13 – The output DataFrame that contains failed checks on multiple columns

How it works...

There are two check levels, **WARNING** and **ERROR**. We only demonstrated the **WARNING** option, but notice that you can implement alerts depending on the check level each data quality check might be assigned to.

The .are_complete() and .are_unique() methods apply data quality checks for each column individually. The resultant DataFrame includes the average number of violations and the pass rate across the input columns.

There's more...

You can use assertion to raise an error if a quality check didn't pass:

```
check = Check(CheckLevel.WARNING, 'Completeness')
result = (
    check
    .is_complete('Type 2')
    .validate(df)
    .select('status')[0,0]=='PASS'
)
assert result
```

The preceding code will raise an assertion error like the following to notify you that the data quality check didn't pass:

```
>>
AssertionError:
```

See also

There are many more data quality tests and checks that you can apply to your DataFrame with cuallee. You can learn more on their GitHub page: https://github.com/canimus/cuallee.

Running unit tests with pytest

pytest is a Python testing framework that allows you to implement all kinds of tests, such as unit tests, integration tests, and end-to-end tests. pytest simplifies your work to develop these tests, as it requires less boilerplate code. If you're already familiar with constructing functions and how to use the `assert` keyword, this will come naturally to you.

A unit test is a crucial step in testing. It verifies the functionality and correctness of each component. It helps identify any errors in your code before the code is put into production.

In this recipe, we'll cover how to run unit tests using pytest in Polars.

Getting ready

You need to have pytest installed. Execute the following command in your terminal to install the library:

```
pip install pytest
```

We need to create the project structure for this recipe. We'll create a `demo.py` file that contains all the transformation logic, and `test_demo.py` under the test folder will contain all the tests that we want to execute using pytest. The folder structure looks like the following:

```
Chapter12/
└── pytest_demo/
    └── demo.py
    └── test/
        └── __init__.py
        └── test_demo.py
```

Let's change the current directory from `Polars-Cookbook` to `pytest_demo`. Run the following code in your terminal:

```
cd Chapter12/pytest_demo
```

Now you're ready to run tests using pytest.

How to do it...

Here's how to run unit tests with pytest in Polars.

1. Define transformations logic in `demo.py`. Let's use functions like the ones we used earlier in this chapter. The following code blocks are defined in one file at https://github.com/PacktPublishing/Polars-Cookbook/blob/main/Chapter12/pytest_demo/demo.py.

    I.   Let's first import the Polars library:

    ```
    import polars as pl
    ```

II.   Define a function to add a rank column:

```
def add_ranks(df):
    return (
        df
        .with_columns(
            pl.col('Attack').rank(method='dense').alias('Atk
Rank'),
            pl.col('Defense').rank(method='dense').
alias('Def Rank'),
            pl.col('Speed').rank(method='dense').alias('Spe
Rank'),
        )
    )
```

III.   Define a function to select necessary columns:

```
def keep_cols(df):
    return (
        df
        .select(
        'Name',
        'Type 1',
        'Type 2',
        'Total',
        'Attack',
        'Defense',
        'Speed',
        pl.col('^*Rank$')
        )
    )
```

IV.   Define a function to filter rows based on Pokemon types:

```
def keep_grass_or_fire(df):
    accepted_types = ['Grass', 'Fire']
    return (
        df
        .filter(
            (pl.col('Type 1').is_in(accepted_types))
            | (pl.col('Type 2').is_in(accepted_types))
        )
    )
```

2. Define tests to run. You can find the whole code in a file at https://github.com/ PacktPublishing/Polars-Cookbook/blob/main/Chapter12/pytest_demo/ test/test_demo.py:

I.   Import the necessary libraries:

```
import pytest
import polars as pl
from cuallee import Check, CheckLevel
from demo import keep_grass_or_fire, keep_cols, add_ranks
```

II. Define a function to read data and apply transformations. Note that pytest fixture is applied to this function:

```
@pytest.fixture(scope='session')
def df():
    df = (
        pl.read_csv('../../data/pokemon.csv')
        .pipe(keep_grass_or_fire)
        .pipe(keep_cols)
        .pipe(add_ranks)
    )
    return df
```

III. Define a test to validate the selected columns:

```
def test_selected_cols(df):
    cols = [
        'Name',
        'Type 1',
        'Type 2',
        'Total',
        'Attack',
        'Defense',
        'Speed',
        'Atk Rank',
        'Def Rank',
        'Spe Rank'
    ]
    assert sorted(df.columns) == sorted(cols)
```

IV. Define a test to validate the accepted types:

```
def test_grass_or_fire_type(df):
    accepted_types = ['Grass', 'Fire']
    filtered_df = (
```

```
            df
            .filter(
                (pl.col('Type 1').is_in(accepted_types))
                | (pl.col('Type 2').is_in(accepted_types))
            )
        )
    assert filtered_df.height == df.height
```

V.   Define a test to validate rank values are within the expected range:

```
def test_rank_range(df):
    bool_df = (
        df
        .select(
            pl.col('^*Rank$').is_between(1, df.height)
        )
        .select(
            pl.all_horizontal(pl.all())
        )
        .select(
            pl.all('^*Rank$')
        )
    )
    assert bool_df.item() is True
```

VI.   Define a test to validate the completeness of rank columns:

```
def test_rank_completeness(df):
    check = Check(CheckLevel.WARNING, 'Completeness')
    cols = ['Atk Rank', 'Def Rank', 'Spe Rank']
    res_df = (
        check
        .are_complete(cols)
        .validate(df)
    )
    assert res_df.select('status').item()=='PASS'
```

3.   Run your tests and check the output:

Make sure that you're in the pytest_demo directory. In your terminal, simply run pytest.

You'll get output like the following:

```
>>
===================== test session starts
=======================
platform darwin -- Python 3.11.3, pytest-8.1.1, pluggy-1.4.0
```

```
rootdir: /Users/Yuki/Desktop/Polars-Cookbook/Chapter12/pytest_
demo
plugins: Faker-19.13.0
collected 4 items


test/test_demo.py
....                                         [100%]


======================== 4 passed in 0.37s
=========================
```

How it works...

When you run your tests with pytest, it searches for and runs files of the test_*.py or *_test.py forms in the current directory and its subdirectories. pytest recognizes the test functions that start with a test_ prefix and automatically runs them all. It's a very simple setup and you can get started quickly.

The df() function that we defined in *step 2* with a decorator is called a fixture. Fixtures are different from test functions in that they're used to define steps or manage dependencies and system states. A good example of using a fixture is when multiple tests depend on the same data or on a network connection. We saw that in our example, wherein we defined @pytest.fixture(scope='session'). You can modify the scope parameter so that you can choose to keep the fixture through the test session, module, and so on. In our example, we defined the scope as session, meaning that the fixture is only called once regardless of how many times it is used by your tests. You only read the data once and use it many times within your session.

Using a data quality testing framework with pytest is a great way to enhance your testing process. You can see how easily you can do that with the assert keyword without modifying the original test code using cuallee.

There's more...

Polars has several useful built-in functions to test the quality of the output DataFrames and Series. Two of such functions are assert_frame_equal and assert_series_equal. You can import them from Polars' testing module.

Let's add another test function to test_demo.py to test the output DataFrame:

```
from polars.testing import import assert_frame_equal

def test_df_equal(df):
    result_df = pl.read_csv('../../data/pytest_expected_output.csv')
    return assert_frame_equal(df, result_df, check_dtypes=False)
```

Let's see how we can run this specific test function only. Run the following code in the terminal:

```
pytest test/test_demo.py::test_df_equal
```

You'll get output like the following:

```
>>
======================== test session starts ========================
platform darwin -- Python 3.11.3, pytest-8.1.1, pluggy-1.4.0
rootdir: /Users/Yuki/Desktop/Polars-Cookbook/Chapter12/pytest_demo
plugins: Faker-19.13.0
collected 1 item

test/test_demo.py .                                         [100%]

======================== 1 passed in 0.33s ========================
```

See also

Please see the following resources to learn more about pytest and Polars' built-in testing functions:

- https://docs.pytest.org/en/7.1.x/contents.html
- https://docs.pola.rs/py-polars/html/reference/testing.html

Index

`packtpub.com`

Subscribe to our online digital library for full access to over 7,000 books and videos, as well as industry leading tools to help you plan your personal development and advance your career. For more information, please visit our website.

Why subscribe?

- Spend less time learning and more time coding with practical eBooks and Videos from over 4,000 industry professionals

- Improve your learning with Skill Plans built especially for you

- Get a free eBook or video every month

- Fully searchable for easy access to vital information

- Copy and paste, print, and bookmark content

Did you know that Packt offers eBook versions of every book published, with PDF and ePub files available? You can upgrade to the eBook version at `packtpub.com` and as a print book customer, you are entitled to a discount on the eBook copy. Get in touch with us at `customercare@packtpub.com` for more details.

At `www.packtpub.com`, you can also read a collection of free technical articles, sign up for a range of free newsletters, and receive exclusive discounts and offers on Packt books and eBooks.

Other Books You May Enjoy

If you enjoyed this book, you may be interested in these other books by Packt:

Getting Started with DuckDB

Simon Aubury, Ned Letcher

ISBN: 978-1-80324-100-5

- Understand the properties and applications of a columnar in-process database
- Use SQL to load, transform, and query a range of data formats
- Discover DuckDB's rich extensions and learn how to apply them
- Use nested data types to model semi-structured data and extract and model JSON data
- Integrate DuckDB into your Python and R analytical workflows
- Effectively leverage DuckDB's convenient SQL enhancements
- Explore the wider ecosystem and pathways for building DuckDB-powered data applications

Data Engineering with Databricks Cookbook

Pulkit Chadha

ISBN: 978-1-83763-335-7

- Perform data loading, ingestion, and processing with Apache Spark
- Discover data transformation techniques and custom user-defined functions (UDFs) in Apache Spark
- Manage and optimize Delta tables with Apache Spark and Delta Lake APIs
- Use Spark Structured Streaming for real-time data processing
- Optimize Apache Spark application and Delta table query performance
- Implement DataOps and DevOps practices on Databricks
- Orchestrate data pipelines with Delta Live Tables and Databricks Workflows
- Implement data governance policies with Unity Catalog

Packt is searching for authors like you

If you're interested in becoming an author for Packt, please visit authors.packtpub.com and apply today. We have worked with thousands of developers and tech professionals, just like you, to help them share their insight with the global tech community. You can make a general application, apply for a specific hot topic that we are recruiting an author for, or submit your own idea.

Share Your Thoughts

Now you've finished *Polars Cookbook*, we'd love to hear your thoughts! Scan the QR code below to go straight to the Amazon review page for this book and share your feedback or leave a review on the site that you purchased it from.

https://packt.link/r/1-805-12115-4

Your review is important to us and the tech community and will help us make sure we're delivering excellent quality content.

Download a free PDF copy of this book

Thanks for purchasing this book!

Do you like to read on the go but are unable to carry your print books everywhere?

Is your eBook purchase not compatible with the device of your choice?

Don't worry, now with every Packt book you get a DRM-free PDF version of that book at no cost.

Read anywhere, any place, on any device. Search, copy, and paste code from your favorite technical books directly into your application.

The perks don't stop there, you can get exclusive access to discounts, newsletters, and great free content in your inbox daily

Follow these simple steps to get the benefits:

1. Scan the QR code or visit the link below

https://packt.link/free-ebook/9781805121152

2. Submit your proof of purchase
3. That's it! We'll send your free PDF and other benefits to your email directly